Atlas of Airborne Fungal Spores in Europe

in Europe

Edited by Siwert Nilsson

Authors: A. Käärik, J. Keller, E. Kiffer
J. Perreau, and O. Reisinger

With 87 Plates and 10 Figures

Springer-Verlag
Berlin Heidelberg New York 1983

Dr. SIWERT NILSSON

Swedish Museum of Natural History
Palynological Laboratory
S-10405 Stockholm, Sweden

ISBN 3-540-11900-0 Springer-Verlag Berlin Heidelberg New York
ISBN 0-387-11900-0 Springer-Verlag New York Heidelberg Berlin

Library of Congress Cataloging in Publication Data. Main entry under title: Atlas of airborne
fungal spores in Europe. Bibliography: p. Includes index. 1. Fungi – Spores – Atlases. 2. Fungi
– Europe – Atlases. 3. Air – Microbiology – Atlases. I. Nilsson, S. (Siwert), 1933 – . II. Käärik,
A. (Aino), 1918 – . QK601.A79   1982   589.2′046′0222   82–16898.

Reproduction of the plates: Gustav Dreher, Württemb. graphische Kunstanstalt GmbH,
Stuttgart.
Cover design: W. Eisenschink, Heddesheim.

Typesetting: Daten- und Lichtsatz-Service, Würzburg
Printing and Bookbinding: Graphischer Betrieb, Konrad Triltsch, Würzburg
2131/3130-543210

Preface

The present atlas is addressed mainly to those who, departing from different spheres of interest, are studying the dispersal of fungus spores in the air, like aerobiologists, plant pathologists, medical mycologists, allergologists, or those interested in spore morphology as a taxonomic tool.

The steadily increasing interest in air pollution problems has also stimulated investigations in the microbiological fraction of air pollutants. During the last decades the study of microbial life in the atmosphere has developed to a special branch of biological sciences – aerobiology of which the first studies are from the end of the last century. Besides pollen grains and spores of higher plants and bacteria, fungal diaspores contribute a constant and substantial part of the airborne microorganisms. Methods for studying airborne spores are described and critically discussed by e.g. Dimmick and Akers (1969), Ingold (1971) and Gregory (1973). The actual content of fungus spores in the air is a result of complicated processes, all of which are influenced by external factors in different ways. The influence of the different meteorological factors on spore formation, liberation, transport and deposition is essentially discussed by Ingold and Gregory.

There are considerable difficulties in identifying detached fungus spores visually. The taxonomy of the fungi is often based on the ontogeny of the spores which cannot be followed by examination of single spores in the air. The conidial development in Deuteromycotina and structure of conidiophores are excellently described and illustrated by Cole and Samson (1979). Among the about 150,000 recognized species of fungi, some common types of spores may occur in a considerable number. In a relatively few cases the spores are recognized to species level; more often the spores may be identified to genus. Thorough knowledge of the local fungus flora or spore cultures may be helpful in identifying the trapped spores.

General spore atlases are few. Apart from Gregory (1973) spores are described and depicted by Bassett et al. (1978).

The present atlas contains illustrations and descriptions of a limited number of fungal diaspores, commonly found in special environments; also common airborne fungi found in natural environments are included with reference to their presence in the air and morphological diversity.

SIWERT NILSSON

Contents

List of Photographers

Introduction

This atlas contains descriptions and illustrations of fungal diaspores of about 90 species from Central and Northern Europe, selected with reference to their relative presence in the air, taxonomic representativity and morphological variability.

Material

Freshly collected spores were generally examined and illustrated. Voucher slides and photos are deposited in the Botanical Institute, University of Neuchâtel, Switzerland (Myxomycota, Basidiomycotina, J. Keller), Department of Forest Products, Swedish University of Agriculture, Uppsala, Sweden (Zygomycotina, Ascomycotina, A. Käärik), Laboratoire de Botanique et de Microbiologie, Université de Nancy, France (Deuteromycotina, material from cultures or natural substrates, E. Kiffer and O. Reisinger) and Museum National d'Histoire Naturelle, Paris, France (Basidiomycotina, Perreau).

Methods

LM (*light microscopy*). – The spores were examined and measured in water (or Melzer's solution for "amyloid" basidiospores).

SEM (*scanning electron microscopy*). – Spores of Myxomycota and Basidiomycotina were fixed in 1% $KMnO_4$ for 1/2 h, dehydrated in acetone, critical point treated, coated with gold and examined in a Philips 200 at 25 kV. Spores of Zygomycotina were treated as above and examined in a Jeol P 15. Spores of Ascomycotina were air-dried, coated with gold and examined in a Jeol P 15. Spores of Deuteromycotina were fixed in 1% OsO_4 for 1 1/2 h, dehydrated in acetone, treated with the critical point method and coated with gold.

TEM (*transmission electron microscopy*). – Spores of Myxomycota and Basidiomycotina were fixed in 1% $KMnO_4$ for 1/2 h, dehydrated in acetone, embedded in epon, sectioned with diamond knife and examined in a Philips 201 at 60 kV. Spores of Zygomycotina and Ascomycotina were fixed in 2% glutaraldehyde in sodium cacodylate buffer and postfixed in 0.2% OsO_4, embedded in Spurr, sectioned and contrast-stained in uranyl acetate and lead citrate. The sections were examined in a Zeiss 10A. Spores of Deuteromycotina were fixed in 2% OsO_4 for 1 1/2 h, dehydrated in ethylalcohol and embedded in epon. The sections were made by means of a diamond knife and contrast-stained with lead citrate.

Terminology

The commonly used terms are listed on pp. 19–22. The concepts and definitions are in accordance with Ainsworth and Bisby's *Dictionary of the Fungi*, 6th ed. (1971), Stearn (1973) and Swartz (1971). The nomenclature follows Martin and Alexopoulos (1969, Myxomycota), Ainsworth et al. (1973, Zygomycotina), Demoulin (1968, Basidiomycotina, Gasteromycetes), Dennis (1978, Ascomycotina), Donk (1956, 1957, 1958, 1960, Basidiomycotina, Hymenomycetes-Aphyllophorales) and Singer (1975, Basidiomycotina, Hymenomycetes-Agaricales).

Spore Descriptions

The descriptions and illustrations are arranged with divisions and subdivisions in taxonomic order (Ainsworth et al. 1973, Ainsworth and Bisby's *Dictionary of the Fungi*, 6th ed., 1971), while orders, families and species are in alphabetical order in the subdivisions.

Illustrations

The species included are illustrated by LM, SEM, and in a few cases TEM. For LM a magnification of ×2000 has been used when feasible.

Financial Support

Laboratoire des Stallergènes, France and Pharmacia Diagnostics, Sweden have provided financial support.

Terminology and Life-Cycles

MYXOMYCOTA (J. Keller and J. Perreau)

The Myxomycota are organisms which are distributed all over the world. During the vegetative state of their life-cycle, they appear as a diploid, multinucleate amoeboid mass of protoplasm, the *plasmodium*. Under specific environmental conditions, the plasmodium develops into one or several *sporocarps* of great morphological diversity. Haploid spores are formed externally, or internally, e.g., in the Myxogastromycetidae (Fig. 1); the majority of them are *encysted spores*. By germination of an encysted spore, one to four naked *myxamoebae* (flagellated swarm cells) are produced; they fuse to a diploid zygote. The zygote is, simply speaking, the beginning of a new plasmodium.

The encysted spores, generally produced in great number, play the most important role in reproduction and are often disseminated by air currents and winds, singly or in cluster.

The morphology of these encysted spores is of high taxonomic significance. The spores are *hyaline* or dark in transmitted light, in mass they appear coloured due to a large scale of pigmentations: from yellow or orange to pink or pinkish grey, red, brick-red, purple or violet to ochraceous or brown. Their shape is generally globose (*Trichia scabra*, Pl. 4) to subglobose

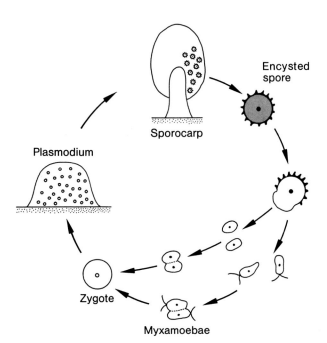

Fig. 1. Life-cycle of the Myxogastromycetidae (Myxomycota)

3

(*Arcyria denudata*, Pl. 2) or ellipsoid (*Trichia floriformis*, Pl. 3), with a diameter of generally 8–10 µm. The wall is thin and its stratification known only for a few species. It seems to consist of three layers (*Lycogala epidendrum*, Pl. 1). The wall is smooth, nearly smooth or, in its outer part, irregularly thickened. The ornamentation can be minutely warted or *echinulate*, rough with irregular or hemispherical processes (*Trichia floriformis*, Pl. 3; *Arcyria denudata*, Pl. 2) or *reticulate* (*Lycogala epidendrum*, Pl. 1; *Trichia scabra*, Pl. 4).

EUMYCOTA
Mastigomycotina and Zygomycotina (A. Käärik)

The Mastigomycotina and Zygomycotina are two large groups of fungi containing about 1500 species, from few-celled, minute organisms to those with conspicuous *thalli* of branched filamentous *hyphae*. Some are strictly parasitic, while others are wholly saprophytic. A large number are aquatic, some amphibious and others wholly terrestrial, among them a number of common, ubiquitous moulds. The assimilatory phase of these fungi is usually coenocytic, i.e. non-septate and multinucleate. In connection with the formation of reproductive structures, septations are commonly formed. In general, both sexual and asexual reproduction occur. The aquatic and amphibious species produce motile spores; the terrestrial species produce non-motile spores, distributed by air or rain. In the Mastigomycotina and Zygomycotina the asexual spores (*sporangiospores*) are produced in an indefinite number in closed *sporangia*, with usually rounded receptacles. Sexual reproduction occurs after the fusion of nuclei from opposite mating types, and this process varies greatly in the different genera of the Mastigomycotina and Zygomycotina. The terrestrial Mastigomycotina and Zygomycotina, including about 3/4 of the total number of species are the only ones of interest in the aerial spore flora.

The largest group of the terrestrial Zygomycotina belongs to the order Mucorales, worldwide in distribution. They are mainly saprophytic on vegetable matter, abundant in soil and on plant debris. Their sporangiospores are produced in large numbers and are often distributed by air. The sexual spores, *zygospores*, are mainly resting spores and are produced after fusion of the nuclei. In the largest family of the order, the Mucoraceae, the mating strains are not morphologically differentiated. In the homothallic species, nuclei of any strain may be mated with nuclei of any other strain. In the heterothallic species, the morphologically similar strains are divided into two

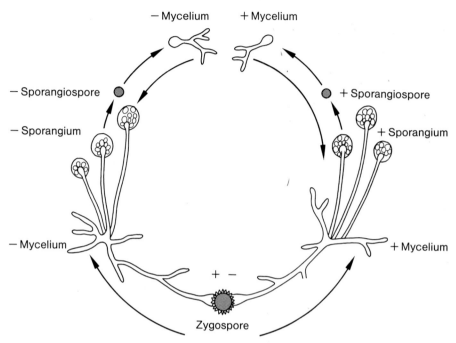

Fig. 2. Life-cycle of the Mucoraceae (Zygomycotina)

opposite mating types, the + and − strains, and nuclear fusion occurs only between the opposite mating types (Fig. 2). In the Mucoraceae, the large and closely related genera *Absidia*, *Mortierella*, *Mucor* and *Rhizopus* are common on plant debris. They are also important, in causing different kinds of storage decay. Their sporangiospores may be distributed by air currents. These sporangiospores are of different dimensions, from 2−3 μm (*Mortierella ramanniana*, Pl. 5) to 10−15 μm (*Rhizopus stolonifer*, Pl. 7); they vary from subglobose to ovoid or angularly ovoid to *pyriform*. In the relatively short-lived sporangiospores the spore wall is well developed and stratified into several layers (*Rhizopus rhizopodiformis*, Pl. 6), often showing an outer ornamentation (*Mortierella*, *Rhizopus*). The spore form and dimensions vary considerably within one species and even in one and the same sporangium. The sporangiospore morphology in the Zygomycotina has taxonomic importance at the species level only.

Ascomycotina (A. Käärik)

About 15,000 species of fungi belong to the Ascomycotina. They show a startling multiplicity and complexity of structural design and an almost infinite variety in pattern of activity. The assimilatory phase of the Ascomycotina consists of a mass of septate hyphae of uninucleate cells. The sexual reproduction is provided by *ascospores*, produced in sac-like specialized terminal cells, the *asci*. The young ascus, after a sexual process, is binucleate until the nuclei fuse to form the diploid nucleus of the primary ascus. Normally, eight ascospores are produced in one ascus, sometimes four or two. The asci may be formed singly, in loose clusters, or arranged in definite fruit bodies, the *ascocarps*, which show an almost endless variety in form and structure. A large number of fungi belonging to the Ascomycotina also have one or more asexual or conidial stage by means of which they propagate and disseminate themselves. Great diversity is also found in the form and dimensions of the *conidia* and in the more or less complex structures for their production. Compared with the ascospores, the conidia are often relatively short-lived, but they may also survive from one season to the next. The life cycle of *Nectria cinnabarina*, producing both ascospores and conidia, is shown in Fig. 3.

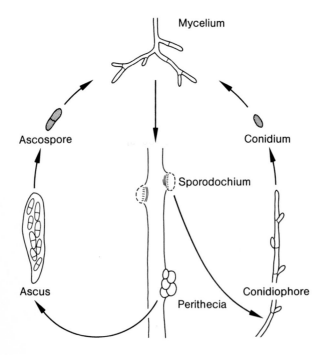

Mycelium

Ascospore

Conidium

Sporodochium

Ascus

Perithecia

Conidiophore

Fig. 3. Life-cycle of *Nectria cinnabarina* (Ascomycotina)

The ascospores and conidia are often distributed by air currents. In some species the ascospores are shot forcibly into the air, in others they are deliberated by the digestion of the ascus wall at maturity. Distribution of ascospores by water, insects, and other agents also occurs. The conidia are often dispersed by air and rain and also by insects and other agents.

The ascospores vary greatly as to dimensions, form and colour. In transmitted light they are hyaline to dark in different shades, from yellowish and greenish to coppery and brown and to the extreme of opaque black. In form they vary from spherical to often ellipsoid or ovoid, sometimes sickle-shaped or *filiform*, with or without a *germ slit* or *germ pores*. They may be one- to many-celled. The cell wall may be smooth or nearly so, or have more or less conspicuous thickenings in form of warts, different kinds of processes, ridges or reticulum. The morphology of the ascospores has taxonomic importance at generic level and sometimes at family level as well.

Of the very numerous species of Ascomycotina some of the large orders, Sphaeriales, Pezizales, Helotiales and Tuberales, which at least locally may be found in great numbers in the air, are included here.

Basidiomycotina (J. Keller and J. Perreau)

Teliomycetes

Uredinales and Ustilaginales (Basidiomycotina, class Teliomycetes) are phytoparasitic fungi without *basidiocarps* and with encysted probasidia called *teliospores*.

Uredinales

The Uredinales or rust fungi, including almost 5000 species, are obligate parasites on wild and cultivated vascular plants, and subsequently of great economic importance. They are autoecious or heteroecious. Their pleomorphic life-cycle is exemplified by Pucciniaceae (Fig. 4).

After germination of a *basidiospore*, the haploid *mycelium* grows on a host A and produces *pycnidia* with *pycnospores*. Through sexual fusion *aecia* with dicaryotic *aeciospores* are formed; these spores are released and distributed by the wind. The germination of an aeciospore results in a dicaryotic mycelium on a host B. *Uredia* with dicaryotic *uredospores* and *telia* with *teliospores*, where each cell is a probasidium, are formed. Caryogamy and meiosis occur in teliospores which produce archaeobasidia with haploid basidiospores borne on *sterigmata*.

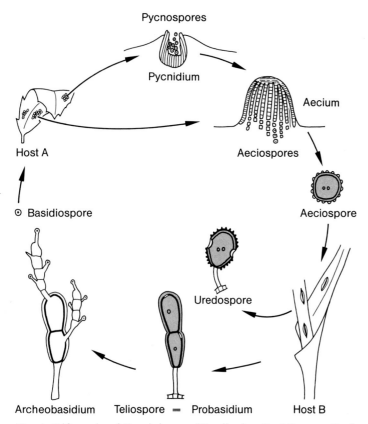

Pycnospores

Pycnidium

Aecium

Host A

Aeciospores

⊙ Basidiospore

Aeciospore

Uredospore

Archeobasidium   Teliospore = Probasidium   Host B

Fig. 4. Life-cycle of Pucciniaceae (Uredinales, Basidiomycotina)

Among the numerous types of spores belonging to the rusts, airborne aeciospores, uredospores, and teliospores are important for dispersal of the species.

Aeciospores are generally hyaline or yellowish in transmitted light and yellow in mass; they are typically single-celled, globose (*Puccinia graminis*, Pl. 18), ellipsoid or polyhedral, (*Cronartium asclepiadeum*, Pl. 16), with a rather large diameter, 15 to 25 µm. The wall is not very thick and more or less completely ornamented with *pustulate* processes or with various, often cylindrical warts, (*Cronartium asclepiadeum*, Pl. 16). The ornamentation seems to be of exosporial origin.

The uredospores vary in colour from yellowish to pale brown in transmitted light and from yellow or orange to rusty brown in mass. They are one-celled on a short *pedicel*, globose (*Phragmidium tuberculatum*, Pl. 17),

8

ovoid (*Puccinia graminis*, Pl. 18), or ellipsoid (*Puccinia triticina*, Pl. 19) generally 20–30 μm in diameter. The wall is thin or thick; the distribution and number of pores are of some taxonomic significance (*Puccinia graminis*, Pl. 18 or *Puccinia triticina*, Pl. 19); the ornamentation is echinulate, *ocellate* (*Puccinia graminis* and *Puccinia triticina*, Pl. 18 and 19), or rarely smooth.

Ustilaginales

The Ustilaginales or smut fungi are parasites on numerous plants, especially cereals and grasses. The teliospores or *chlamydospores*, to be regarded as probasidia, are formed in balls which develop in the host tissues and break up easily at maturity. Then, the chlamydospores are liberated singly or in cluster, and carried away by wind or water, infesting new areas.

The germination of a chlamydospore (Fig. 5) results in a septate promycelium with *sporidia* corresponding to haploid basidiospores, which sometimes can bud conidia. A dicaryotic mycelium is formed by fusion of basidiospores or conidia. Specific hosts are invaded. Chlamydospores are pale brown in colour, sometimes violaceous in transmitted light, darker, brown or purplish in mass. The shape is generally subglobose to ellipsoid or somewhat polyhedral (*Tilletia controversa*, Pl. 20), with a diameter ranging from 4–10 μm (*Ustilago avenae*, Pl. 21) to 18–24 μm (*Tilletia controversa*, Pl. 20).

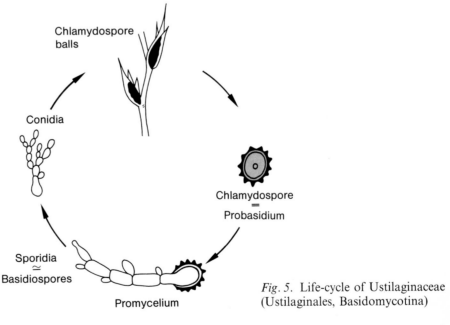

Chlamydospore balls

Conidia

Chlamydospore = Probasidium

Sporidia ≃ Basidiospores

Promycelium

Fig. 5. Life-cycle of Ustilaginaceae (Ustilaginales, Basidomycotina)

In many species the chlamydospores are surrounded by sterile translucent cells. The wall is often relatively thick, smooth or variously ornamented, depending on which part of the spore is observed; verrucose (*Ustilago avenae*, Pl. 21; *Ustilago maydis*, Pl. 22), echinulate, reticulate (*Tilletia controversa*, Pl. 20; *Ustilago tragopogonis-pratensis*, Pl. 23). It consists of a smooth *episporium*, a smooth or irregularly thickened *exosporium*, forming the ornamentation, and a mucilaginous *perisporium*.

Hymenomycetes (Holobasidiomycetidae)

The Holobasidiomycetidae are characterized by aggregate mycelial structures and fruit bodies of extremely varying morphology of taxonomic significance. The fertile part of the fruitbody – the basidiocarp – is the *hymenium* where caryogamy of dicaryons and meiosis occur in each *basidium* which generally produces four basidiospores. The germination of a basidiospore provides a haploid mycelium; fusion of two such mycelia results in a dicaryotic mycelium where plasmogamy occurs but not caryogamy. The dicaryotic mycelium is able to produce fruitbodies (Fig. 6).

Many different kinds of basidiospores are frequently found among airborne organisms. In the Hymenomycetes, the basidiospores occur as *ballistospores* (not in the Gasteromycetes).

The colour of basidiospores in mass – white, cream, yellow, pink, ochraceous, ferrugineous, brown, black, with purplish or greenish tints – attracts taxonomic interest. In transmitted light, the basidiospores appear to be hyaline or subhyaline or with pale colours, e.g., reddish, incarnate, yellowish, brownish, etc.

Mature basidiospores have a shape that varies between taxonomic groups; they are unicellular, bilateral, in lateral view showing a proximal part with *hilar appendix*, an *abaxial* part and an *adaxial* part with a supra-appendicular depression. The shape is globulose (*Amanita fulva*, Pl. 25; *Lactarius lignyotus*, Pl. 37; *Calvatia excipuliformis*, Pl. 60), subglobose (*Tricholomopsis platyphylla*, Pl. 48; *Hydnum repandum*, Pl. 55), ovoid (*Inocybe queletii*, Pl. 30; *Strobilomyces floccopus*, Pl. 43), ellipsoid (*Naematoloma sublateritium*, Pl. 44; *Russula puellaris*, Pl. 42), amygdaliform (*Cortinarius praestans*, Pl. 28), *fusiform* (*Chroogomphus rutilus*, Pl. 33), *allantoid* (*Peniophora quercina*, Pl. 52), cylindraceous, polyhedral (*Rhodophyllus hirtipes*, Pl. 35) or *gibbous* (*Inocybe subcarpta*, Pl. 31; *Phylacteria terrestris*, Pl. 58).

The spore wall is fairly thick and consists of four fundamental layers: *ectosporium*, perisporium/exosporium, episporium and *endosporium*. Some variations as to hilar appendix are noted, with hilar scar and sometimes with

10

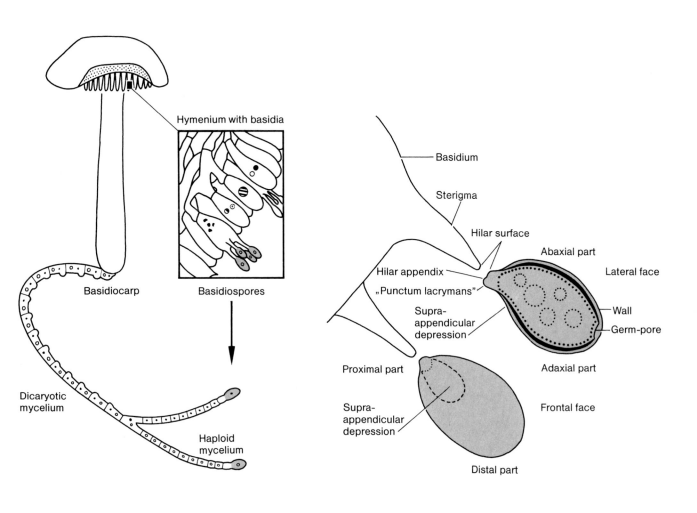

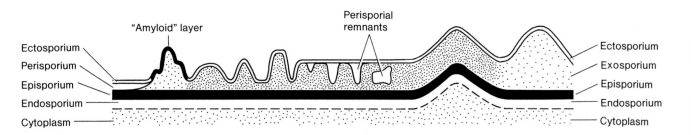

Fig. 6. Life-cycle of Holobasidiomycetidae (Basidiomycotina) and wall stratification

11

remnants of a sterigma wall, and the presence of a germ pore distally (*Naematoloma sublateritium*, Pl. 44). If the shape of the spore is related to the episporium, the real ornamentation is due to irregularities in the exosporium thickness elaborated in the perisporium.

The basidiospores are smooth (*Cantharellus cibarius*, Pl. 50), smooth with a thick brown-pigmented exosporium (*Agaricus bitorquis*, Pl. 24; *Boletus edulis*, Pl. 26; *Inocybe queletii*, Pl. 30), smooth but gibbous (*Inocybe subcarpta*, Pl. 31) or smooth with an "*amyloid*" outer layer (*Amanita fulva*, Pl. 25). They are verrucose (*Rozites caperata*, Pl. 32), pustulate (*Lepista sordida*, Pl. 45), echinulate, with conical warts (*Lycoperdon perlatum*, Pl. 61), cristulate, *striate*, ribbed, *sulcate* (*Clitopilus prunulus*, Pl. 34), *foveate* (*Ganoderma applanatum*, Pl. 53), *alveolate* or reticulate (*Strobilomyces floccopus*, Pl. 43). The ornamentation is *tuberculose* and covered by an "amyloid" layer in, e.g., *Russula* and *Lactarius* (*Lactarius lignyotus*, Pl. 37; *Russula laurocerasi* v. *fragrans*, Pl. 41).

Deuteromycotina (E. Kiffer and O. Reisinger)

Spores (= conidia) in the Deuteromycotina (Fungi Imperfecti) are of vegetative origin. Their formation is not linked with fusion followed by meiosis that is characteristic of the sexual phenomena in the Asco- and Basidiomycotina. Therefore a conidium might be defined as an asexual, exogenous, non-motile spore produced during the life-cycle of the Ascomycotina, Basidiomycotina or fungi related to them by septal structure. Conidia are not simple, undifferentiated hyphal fragments, they exhibit a morphological diversity according to their production, to their degree of maturity, and to their mode of liberation (e.g., thickening of the wall, melanization, reserve accumulation, scars at the attachment points etc). Such characteristics are often useful for specific, generic or suprageneric identifications. They are, however, insufficient for an accurate taxonomic determination, which requires a highly trained specialist. Therefore, for a systematic inventory of airborne spores in a certain environment, it would be necessary to possess a good knowledge of the fungal flora of this particular environment. An additional problem comes from fungi which produce two totally different kinds of spores.

Concepts in Deuteromycotina Systematics

Part of the yeasts, sterile mycelia, and all fungi with regularly septate hyphae but without reproductive organs are grouped among the Deuteromycotina

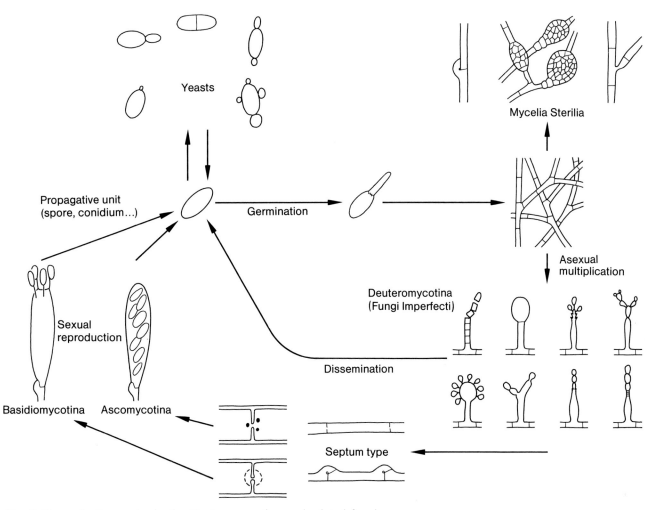

Yeasts

Mycelia Sterilia

Propagative unit
(spore, conidium...)

Germination

Asexual
multiplication

Sexual
reproduction

Deuteromycotina
(Fungi Imperfecti)

Dissemination

Basidiomycotina   Ascomycotina

Septum type

Fig. 7. Reproductive cycles in the Deuteromycotina and related fungi

which do not form a natural systematic unit. The asexual stages of Ascomycotina and Basidiomycotina may also be included (Fig. 7).

The taxonomy of this artificial group has been studied and revised extensively, especially in the last 30 years. Classification of this group may be based on: (1) general morphology, colour and grouping of fruiting structures (Fig. 8); at a lower taxonomic level, shape, colour, and septation of conidia are also used (Fig. 9). (2) conidium ontogeny (Fig. 10). The latter gives more information for identification of isolated conidia with reference to relationships between conidiogenous cell and conidium (= types of conidial ontogeny), including primordium initiation, conidium maturation, conidium separation.

By using conidium ontogeny it is possible to distinguish seven types of conidia (Fig. 10).

It is preferable to use concepts and terms from different nomenclatural systems which allow the best characterization of conidia. It is especially useful for practical purposes when identifying spores separated from their conidiophore.

Taxonomic Significance of Conidia

The conidia of the Deuteromycotina possess a number of features which may provide important clues at various taxonomic levels.

1. Colour, Number of Septa, Shape

Conidia may be hyaline, more or less melanized or brightly coloured (in the latter case green is the most frequent colour). The common spore types are collocated in Fig. 9 where septation and shape of conidia are also taken into account.

2. Wall Structure

Various types of conidia are distinguished with reference to number of wall layers. This has been particularly useful among the phragmospores. Mangenot and Reisinger (1976) proposed to group them with the amerospores into haplothecate (= euseptate) and diplothecate (= distoseptate and intermediary) conidia. In order to recognize distoseptate conidia, Luttrell (1963, 1964) proposed a technique by which young spores are crushed between a microscope slide and a cover slip: haplothecate conidia break into pieces, while diplothecate ones liberate internal loci surrounded by the internal, hyaline layer of the conidial as wall.

14

Sterile mycelium		Agonomycetaceae	Agonomycetales	Hyphomycetes
Separate conidiophores	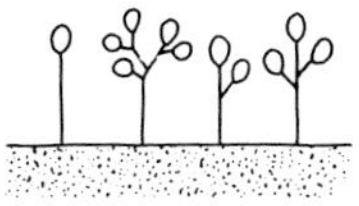	Dematiaceae darkly pigmented / Mucedinaceae hyaline or brightly coloured	Hyphomycetales	
Synnemata (coremia)	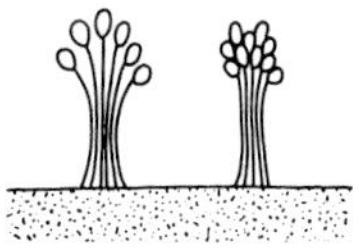	Stilbaceae	Stilbellales	
Cushion-shaped stromata: sporodochia	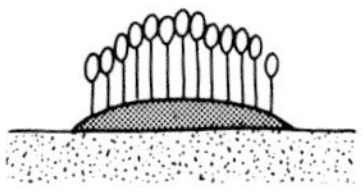	Tuberculariaceae	Tuberculariales	
Thin, immersed stromata: acervuli	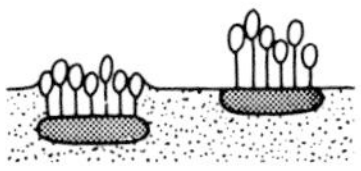	Melanconiaceae	Melanconiales	Coelomycetes
Complete pycnidia	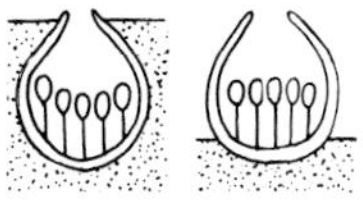	Sphaerioidaceae darkly pigmented / Nectrioidaceae hyaline or brightly coloured	Sphaeropsidales	
Incomplete pycnidia	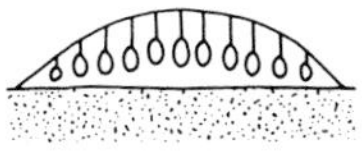	Leptostromataceae		
Cup-shaped pseudopycnidia	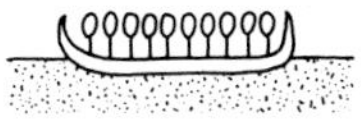	Excipulaceae		

Fig. 8. Classification of the Deuteromycotina based on morphology, colour and grouping of the fruiting structures (Saccardo 1886)

Morphology	Definition	Spore types	Colour	
			Hyaline or brightly coloured	Brown, black or darkly pigmented
 $0\ 0\beta\,l\,\partial\,\bigcirc$	Unicellular, ovoid, globose, cylindrical, allantoid…	Amerospores	Hyalospores	Phaeospores
	Bicellular, ovoid, pyriform, elongate…	Didymospores	Hyalodidymospores	Phaeodidymospores
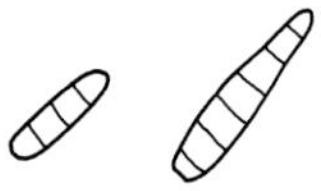	Pluricellular with transverse septa, elongate, fusiform…	Phragmospores	Hyalophragmospores	Phaeophragmospores
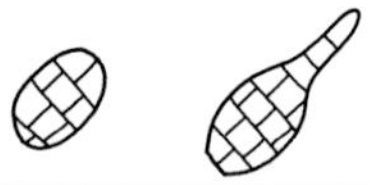	Muriform with transverse and longitudinal septa	Dictyospores	Hyalodictyospores	Phaeodictyospores
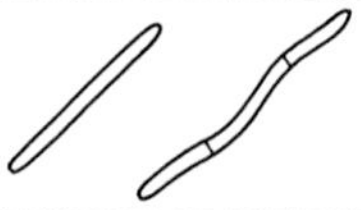	Uni- or pluricellular, acicular, filiform…	Scolecospores	Hyaloscolecospores	Phaeoscolecospores
	Uni- or pluricellular, spiral- or helicoid	Helicospores	Hyalohelicospores	Phaeohelicospores
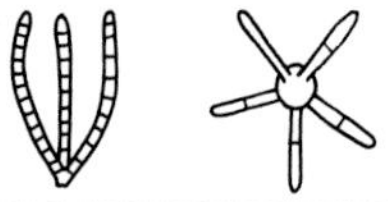	Uni- or pluricellular, star- or cross-shaped, variously branched	Staurospores	Hyalostaurospores	Phaeostaurospores

Fig. 9. Spore types according to shape, septation and colour

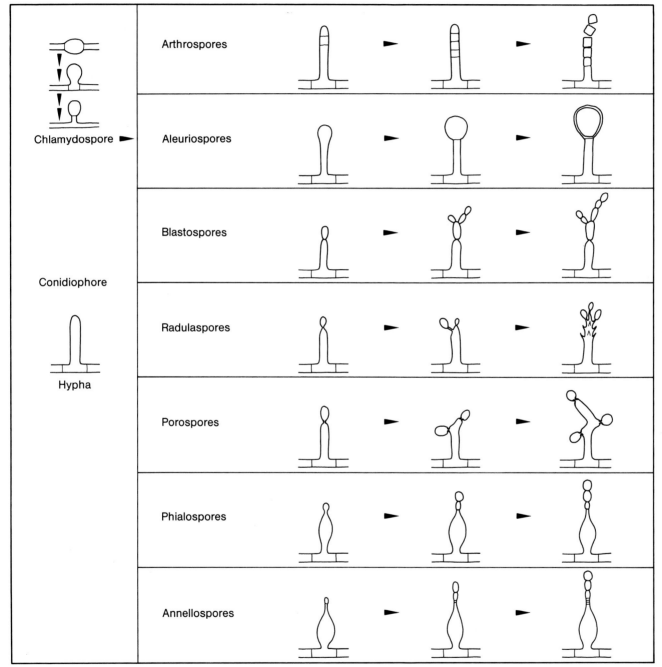

Fig. 10. Spore types according to ontogeny

17

3. Preformed Germinative Structures

These structures are differentiated zones of the conidial wall, with a less resistant area where the germ tube will emerge. They are visible as pores or slits with reduced melanization. Equatorial lines in bivalve conidia (*Nigrospora*, *Arthrinium*) are comparable to germ slits.

4. Ornamentation

The conidial surface may be smooth or ornamented. In the latter case the ornamentation is very variable, with minute echinulations to coarse warts. Warts are limited by a thin layer which encloses a vesicle with no structured content which collapses during sample preparation for SEM (Reisinger et al. 1977). Other surface ornamentation consists of rigid spines and many intermediates between warts and spines.

5. Hilum and Conidial Scars

Fungal spores are either endogenous (ascospores, sporangiospores) or exogenous (basidiospores, conidia). The latter are recognised by the presence of a hilum, which is the trace of the insertion point on the mother cell. If the conidium is solitary there is only one hilum. This is not sufficient to distinguish conidia from basidiospores; the latter usually possess a rather characteristic outline and dissymetry. When conidia form chains, traces of the insertion are usually visible. Thus in the basipetal, unbranched chains of *phialospores* and *annellospores*, conidia may bear two scars, one basal and one apical. In *blastospores* and *porospores* with acropetal, branched chains, each conidium bears one basal hilum but possibly two or more apical scars.

However, some conidia lack visible insertion scars, e.g., the phialoconidia produced in mucous heads. The hilum may be an important feature for determining the origin of conidia; especially in the porospores, the hilum is usually quite visible as an annular melanized thickening around a clear, minute channel.

A wide hilum and conidial scar is usually characteristic of the *aleuriospores*, *arthrospores*, endogenous phialospores and annellospores. In the always solitary *radulaspores*, the hilum is usually a pointed denticle. Blastospores often possess intermediate and rarely characteristic types, although multiple scars mentioned above can be found in this group. Finally, the hilum may bear wall debris, e.g., in aleuriospores and radulaspores.

Glossary of Terms

Abaxial	Away from the axis or central line.
Acervulus (pl. acervuli)	A cushion-shaped, immersed, fertile stroma producing conidia.
Acicular	Needle-shaped.
Adaxial	Towards the axis or the centre.
Aeciospore	Spore of Uredinales produced in an aecium.
Aleuriospore	A conidium developed from the blown-out end of a sporogenous cell or hyphal branch.
Allantoid	Sausage-like.
Alveolate	Honeycombed.
Amygdaliform	Almond-like.
Amyloid	Resembling starch, i.e., giving a dark blue or violaceous reaction to iodine.
Annellophore	A fertile cell with apical annellations resulting from percurrent proliferations accompanying the formation of successive conidia.
Annellospore	A spore borne on an annellophore.
Arthrospore	A spore resulting from the breaking up of a hypha into separate cells.
Ascocarp	The ascus-bearing organ of an ascomycete.
Ascospore	A spore of Ascomycotina produced in an ascus by "free cell formation".
Ascus (pl. asci)	A sac-like cell of the perfect state of Ascomycotina, in which ascospores, generally 8, are produced.
Ballistospore	A forcibly abjected basidiospore.
Basidiocarp	Basidium-bearing organ of a basidiomycete.
Basidiospore	A spore of Basidiomycotina developed on a sterigma of a basidium.
Basidium (pl. basidia)	The organ in Basidiomycotina bearing basidiospores.
Blastospore	A conidium which has been budded off the fertile cell by a narrow isthmus.
Chlamydospore	A thick-walled, reserve-filled, asexual spore. A chlamydospore results from the transformation of part of a hypha. The term has sometimes been extended to other types of spores, (e.g., aleuriospores in Deuteromycotina, teliospores in Ustilaginales).
Conidial scar	Attachment point of a conidium on the fertile cell. Such a scar may exist on a conidiophore or on a fertile conidium which has produced other conidia towards the apex.

Conidiophore	A simple or branched, more or less modified hypha on which conidia are produced.
Conidium (pl. conidia)	An asexual, exogenous, non-mobile spore produced during the life cycle of Ascomycotina, Basidiomycotina and typically Deuteromycotina.
Cristate	Crested.
Distal	Remote from the place of attachment of a spore.
Echinulate	Ornamented with very small prickles (echinulae).
Ectosporium	Outer layer of the basidiospore wall.
Ellipsoid	Elliptical in optical section.
Encysted spore	Kind of fungal spore, especially among the Myxogastromycetidae.
Endosporium	Inner layer of the basidiospore wall.
Episporium	Fibrillous layer of the basidiospore wall, around the endosporium.
Equatorial diameter	The diameter of a spore measured in the equatorial plane.
Equatorial view	Side view of radially symmetrical spores.
Exosporium	Electron-dense or translucent, irregular or smooth layer of the basidiospore wall forming a transition into the perisporium, above the episporium.
Filiform	Thread-like.
Foveate	Pitted.
Fusiform	Spindle-like, tapering towards the ends.
Germ pore	More or less circular differentiation of the spore wall, often exit for the germ tube.
Germ slit	Same as above but elongate instead of circular. Present only in amerospores.
Gibbous	Having a central swelling.
Globose, globulose	Spherical or nearly so.
Hilar appendix	Relatively short pedicel at the proximal part of the basidiospore.
Hilum (pl. hila)	A mark or scar on a spore at the point of attachment to a conidiophore or sterigma.
Hyaline	Transparent or colourless under the microscope.
Hymenium (pl. hymenia)	Spore-bearing layer of a fruit body.
Hypha (pl. hyphae)	One of the filaments of a mycelium.
Lateral view	Side view of bilateral spores.

Muriform	Having transverse and longitudinal septa.
Mycelium (pl. mycelia)	The thallus of a fungus; a mass of hyphae.
Myxamoeba (pl. myxamoebae)	A zoospore after becoming ameoba-like.
Ocellate	With a centre of one colour surrounded by a broad ring of another.
Ovoid	Resembling an egg.
Pedicel	A small proximal stalk.
Perisporium	Layer of the basidiospore wall below the ectosporium.
Perithecium (pl. perithecia)	The subglobose or flask-like ascocarp of the Pyrenomycetes (Ascomycotina).
Phialide	A short, flask-shaped fertile cell from the apex of which conidia are abstricted in basipetal succession.
Phialospore	Spore formed successively to produce chains or spore heads on phialides.
Plasmodium (pl. plasmodia)	A multinucleate, motile mass of protoplasm.
Polar axis	A perpendicular line connecting the poles of a spore.
Polar view	A spore viewed with one of the poles exactly uppermost.
Porospore (tretoconidium)	A conidium produced through a minute pore in the wall of the fertile cell. The latter and the conidium are always melanized and typically possess at the hilum and conidial scar a circular, dark thickening surrounding a clear channel.
Proximal	Near the place of the attachment of a spore.
Pustulate	With a small elevation or spot resembling a blister (pustule).
Pycnospore	A spore formed inside a pycnidium.
Pyriform	Pear-like.
Radulaspore	Solitary conidium, appearing on a narrow isthmus, usually resulting after secession in a pointed conidial scar and hilum.
Reticulate	With a network (reticulum, pl. reticula).
Spinose	With narrow, sharply pointed processes (spina, pl. spinae).
Sporangiospore	Asexual spore produced in a sporangium.
Sporangium (pl. sporangia)	An organ producing endogenous asexual spores (esp. in Myxomycota, Mastigomycotina and Zygomycotina.

Sporidium (pl. sporidia)	A basidiospore of the Ustilaginales.
Sporocarp	A spore-producing organ.
Sporodochium (pl. sporodochia)	An external mass of conidiophores tightly placed upon a stroma or mass of hyphae.
Sterigma (pl. sterigmata)	A spore-bearing process from a basidium.
Striate	With ridges.
Stroma (pl. stromata)	A mass or matrix of vegetative hyphae.
Sulcate	Grooved.
Synnema (pl. synnemata)	A more or less compacted group of erect and sometimes fused conidiophores bearing conidia at the apex only.
Teliospore	Kind of spore among the Uredinales, produced in telia (teleutosori).
Thallus (pl. thalli)	A vegetative body of thallophyte.
Tuberculose	With rounded wart-like processes.
Uredospore	Kind of fungal spore among the Uredinales, produced in uredia (uredosori).
Verrucose	With warts (verruca, pl. verrucae).
Zygospore	Resting spore resulting from the conjugation of two isogametes.

Plates

1   *Lycogala epidendrum* (L.) Fries

Myxomycota · Liceales · Reticulariaceae

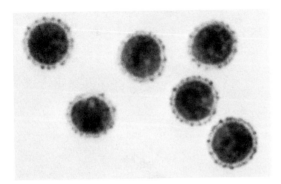

Spherical, reticulate, encysted spores. LM × 2000

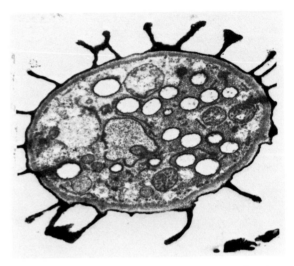

Thin section of an encysted spore with a threelay-ered wall. TEM × 11,000

Spherical encysted spores with reticulate ornamentation and polymorphic meshes. SEM × 11,000 (*to the left*), × 8000 (*to the right*)

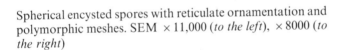

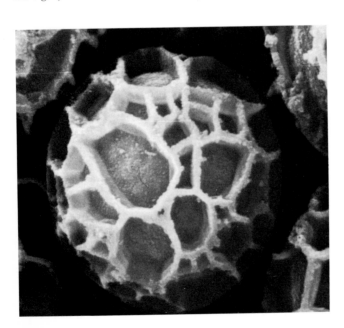

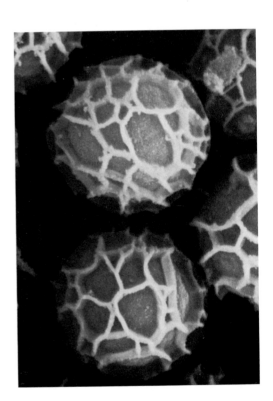

2 *Arcyria denudata* (L.) Wettst.

Myxomycota · Trichiales · Trichiaceae

A subspherical, encysted spore in equatorial view. LM ×2000

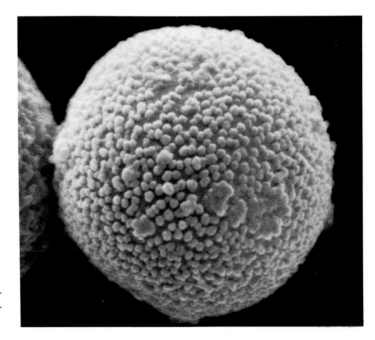

A globose, densely warty encysted spore; some tuberculose processes are scattered among numerous subhemispherical warts. SEM ×10,000

3 *Trichia floriformis* (Schw.) G. Lister

Myxomycota · Trichiales · Trichiaceae

A broadly elliptic encysted spore. LM ×2000

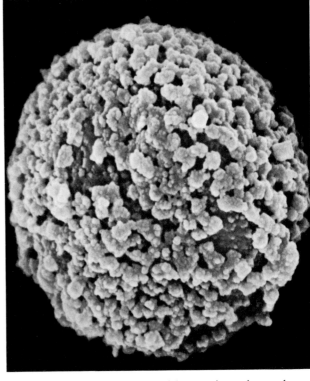

A spherical encysted spore with very irregular scabrous processes. SEM ×8000

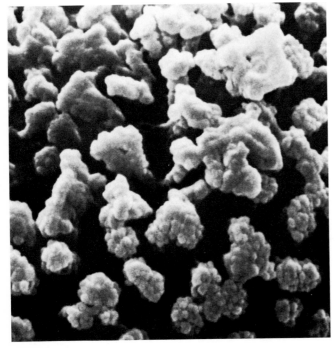

Detail of spore surface showing irregularly nodulose tubercles. SEM ×24,000

4 Trichia scabra Rostrup

Myxomycota · Trichiales · Trichiaceae

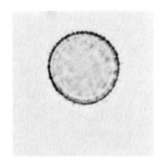

A spherical, encysted spore. LM × 2000

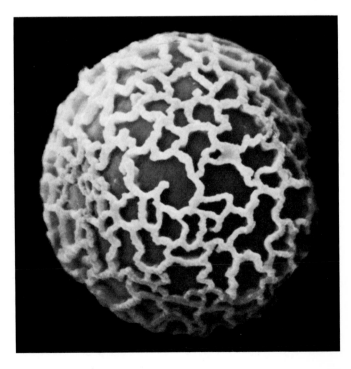

A reticulate encysted spore with a raised festooned network and irregularly shaped interspaces. SEM × 8000

5   *Mortierella ramanniana* (Moeller) Linnemann

Eumycota · Zygomycotina · Mucorales · Mortierellaceae

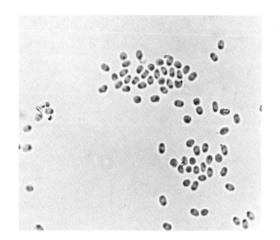

A group of spherical to broadly elliptical minute sporangiospores. LM × 1000

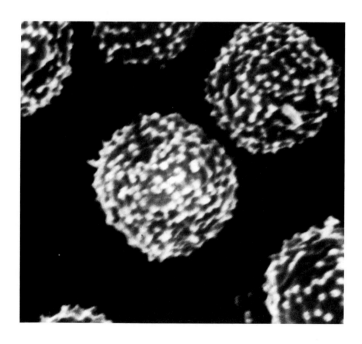

A group of sporangiospores showing spore wall ornamentation. Dense warts and spines in a net-like configuration. SEM × 15,000

28

6 *Rhizopus rhizopodiformis* (Cohn) Zopf

Eumycota · Zygomycotina · Mucorales · Mucoraceae

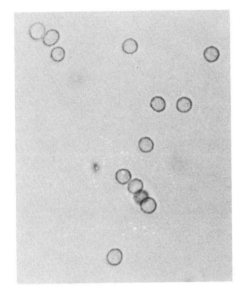

A group of hyaline spiny sporangiospores. LM ×850

Details of the exosporial ornamentation of a sporangiospore. SEM ×20,000

A sporangiospore showing exosporial ornamentation with blunt spinules on ridges. SEM ×12,000

Section of a sporangiospore, showing an electron dense exosporium, an electron transparent inner wall and cytoplasm with organelles. TEM ×10,000

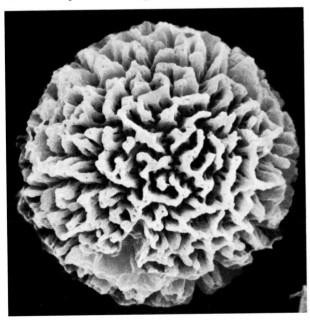

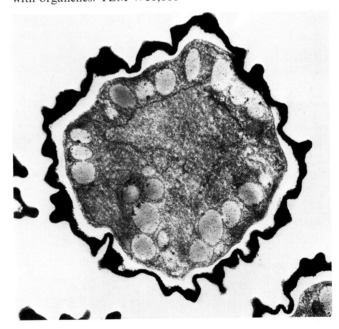

7 *R. stolonifer* (Ehrenb. ex Fr.) Lind

Eumycota · Zygomycotina · Mucorales · Mucoraceae

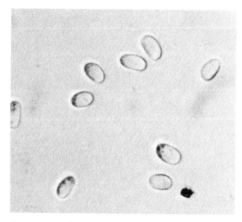

A group of sporangiospores in lactophenol with cotton blue. No wall structure is seen in fresh, water-swollen spores. Sporangiospores are of variable size. LM ×850

A group of sporangiospores in dry condition, showing longitudinal ridges. After swelling in water the ridges more or less disappear. SEM ×1500

A sporangiospore in dry condition. Conspicuous longitudinal exosporial ridges and between them irregular small warts. SEM ×18,500

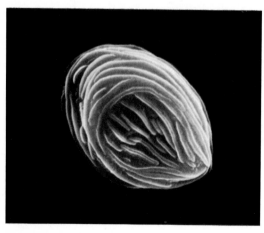

Section of a young ascospore showing a thick crust-like ▷ outer wall, a thin fibrous inner wall and a large vacuole. TEM ×10,000

Detail of a section of an ascospore wall, showing the thick ▷▷ electron-dense fibrous outer wall, an electron transparent thick fibrous inner wall consisting of several layers and an electron dense plasma membran delimiting the cytoplasm with its organelles. TEM ×47,000

8   *Bulgaria inquinans* Pers. ex Fr.

Eumycota · Ascomycotina · Helotiales · Helotiaceae

A group of ascospores. Three dark brown
opaque and two hyaline spores.
LM × 1000

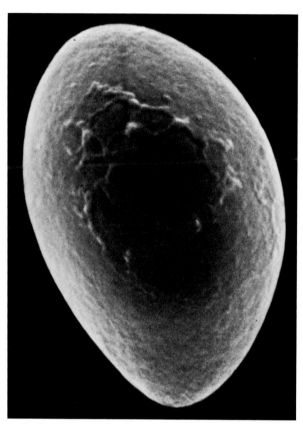

A dark ascospore in dorsal view. Exosporium smooth,
fibrous. SEM × 10,000

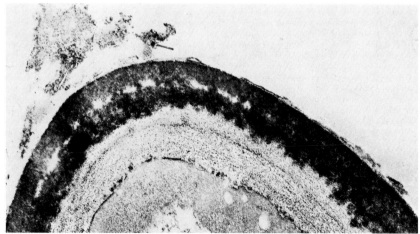

9 *Helvella crispa* Fr.

Eumycota · Ascomycotina · Pezizales · Helvellaceae

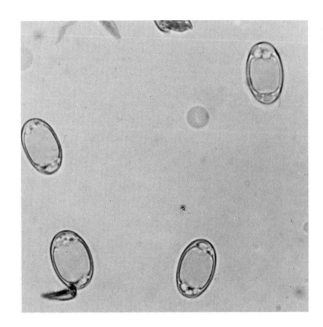

A group of broadly elliptical ascospores with one large central oil drop. LM × 1000

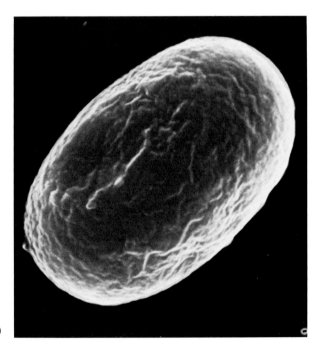

An ascospore with a coarse net-like exosporium. SEM × 4500

10 *Rhizina undulata* Fr.

Eumycota · Ascomycotina · Pezizales · Helvellaceae

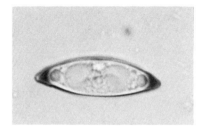

A fusiform apiculate ascospore with several oil drops. In lactophenol with cotton blue. LM × 1000

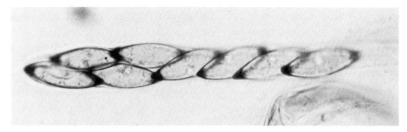

Eight ascospores in an ascus. Ascus wall hyaline, scarcely visible. In lactophenol with cotton blue. LM × 700

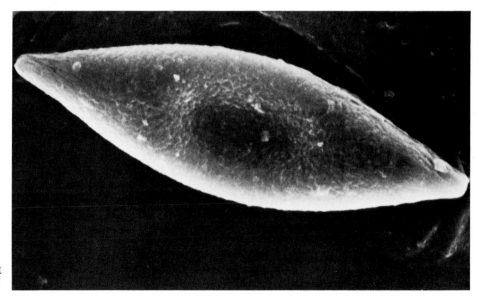

An ascospore with a somewhat coarse exosporium. SEM × 4500

11 *Otidea onotica* (Pers.) Fuckel

Eumycota · Ascomycotina · Pezizales · Pezizaceae

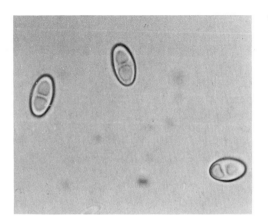

Three broadly elliptical ascospores with two large oil drops. LM × 1000

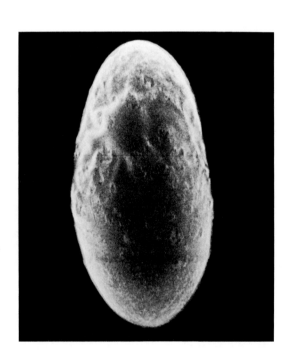

An ascospore, showing a rather smooth exosporium with scattered low warts. SEM × 6000

12 *Elaphomyces granulatus* Fr.

Eumycota · Ascomycotina · Plectascales · Elaphomycetaceae

Three ascospores in low focus. Spores very dark brown, opaque, covered with irregular flattened warts. LM ×1000

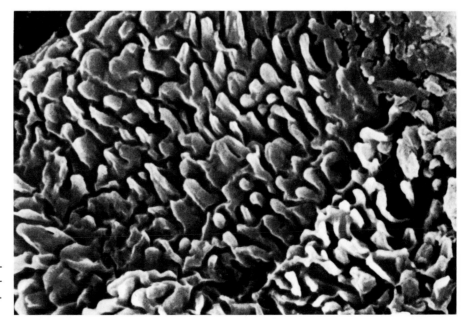

Detail of the ascospore wall structure. Exosporium with a dense irregularly verrucoid ornamentation. SEM ×7500

13 *Chaetomium globosum* Kunze ex Fr.

Eumycota · Ascomycotina · Sphaeriales · Melanosporaceae

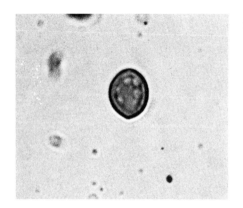

An ascospore, apiculate on both ends, dark olive-brown, with smooth wall. LM ×1000

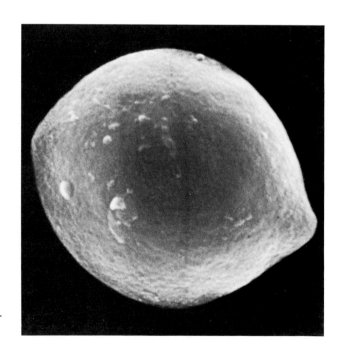

A lemon shaped ascospore showing a fairly smooth exosporium, with a few irregular flattened warts. SEM ×6500

14 Nectria cinnabarina (Tode ex Fr.) Fr.

Eumycota · Ascomycotina · Sphaeriales · Nectriaceae

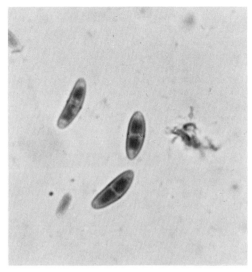

A group of slightly fusiform hyaline asco-spores with a median transverse septum. In lactophenol and cotton blue. LM × 600

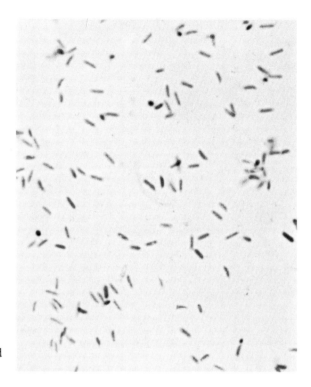

A group of small cylindrical hyaline conidia. In lactophenol and cotton blue. LM × 600

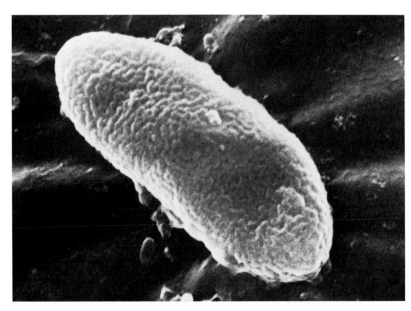

An ascospore, showing a rough, low warty structure of the exosporium. SEM × 5000

15 *Ustulina deusta* (Fr.) Petrak

Eumycota · Ascomycotina · Sphaeriales · Sphaeriaceae

Three fusiform, dark brown, opaque ascospores, with one side flattened. LM ×700

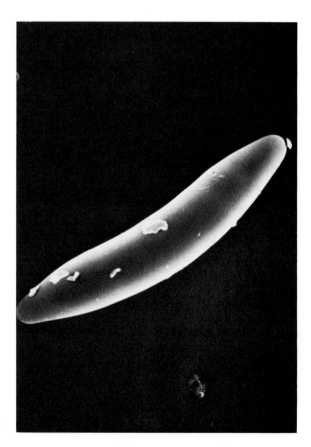

An ascospore showing a very smooth exosporium. SEM ×3000

16 *Cronartium asclepiadeum* (Willd.) Fries

Eumycota · Basidiomycotina · Teliomycetes · Uredinales · Melampsoraceae

Aeciospores in different foci. To the left, an aeciospore in high focus, to the right, another one in optical cross-section. LM × 2000

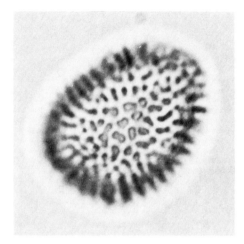
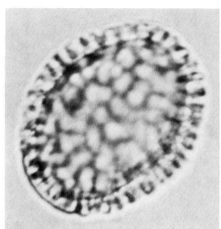

An aeciospore with densely crowded cylindrical pillars. SEM × 3000

Detail of the stratified pillars with irregularly nodulose top. SEM × 18,000

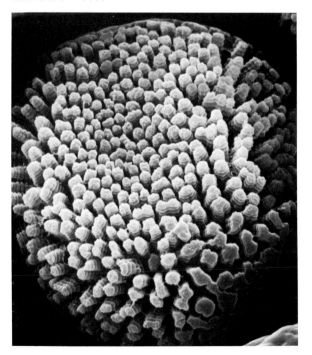

17 *Phragmidium tuberculatum* J. Mueller

Eumycota · Basidiomycotina · Teliomycetes · Uredinales · Pucciniaceae

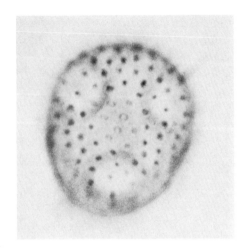
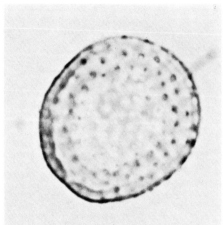

Uredospores in different foci. To the left an uredospore in high focus, the other in optical cross-section. LM ×2000

Uredospores with distinct, regularly arranged conical spines and germ-pores. SEM ×1500

40

17 *Phragmidium tuberculatum* J. Mueller

Eumycota · Basidiomycotina · Teliomycetes · Uredinales · Pucciniaceae

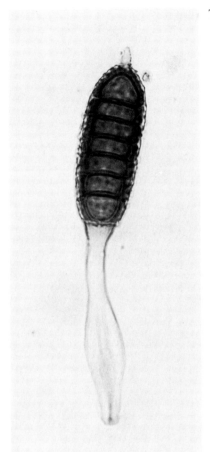

Teliospore with seven cells, a hyaline distal appendice and a long pedicel. LM × 2000

Teliospores ornamented with irregular verrucose tubercles and with enlarged pedicel at the base. SEM × 2700

18 *Puccinia graminis* Persoon

Eumycota · Basidiomycotina · Teliomycetes · Uredinales · Pucciniaceae

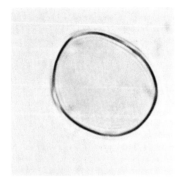

An aeciospore. LM ×2000

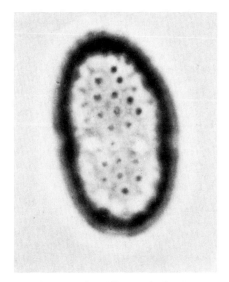
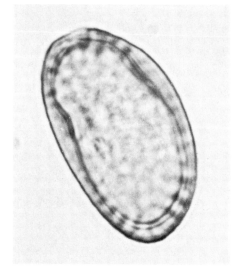

Uredospores in different foci. The uredospore on the left in high focus, the other in optical cross-section. LM ×2000

An aeciospore almost completely covered with numerous warts. SEM ×4000

Detail of uredospore surface showing conical spines, each of them surrounded at the base by a torus. SEM ×4000

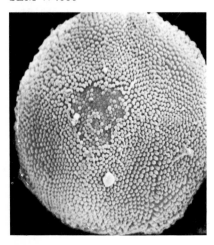
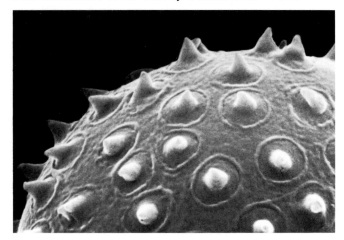

19 *Puccinia triticina* Eriksson

Eumycota · Basidiomycotina · Teliomycetes · Uredinales · Pucciniaceae

Uredospores in different foci. The uredospore to the left in high focus, the other in optical cross-section. LM × 2000

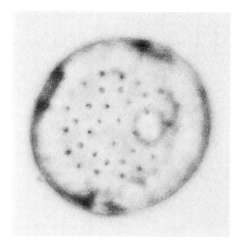 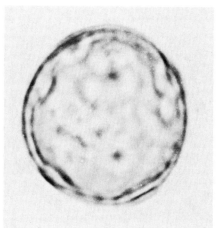

An uredospore covered with numerous conical spines surrounded by a ring-shaped area, and germ-pores. SEM × 4000

Detail of the uredospore ornamentation showing regularly arranged, conical spines surrounded by a torus. SEM × 8000

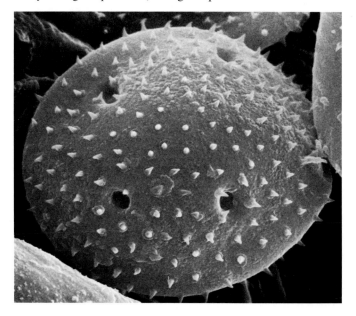 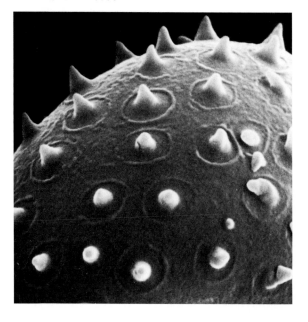

20   *Tilletia controversa* Kühn

Eumycota · Basidiomycotina · Teliomycetes · Ustilaginales · Tilletiaceae

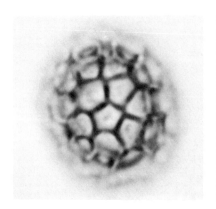 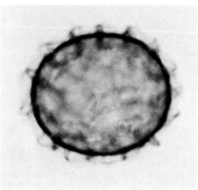

Subglobose chlamydospores in different foci. The chlamydospore to the left in high focus, the other in optical cross-section. LM × 2000

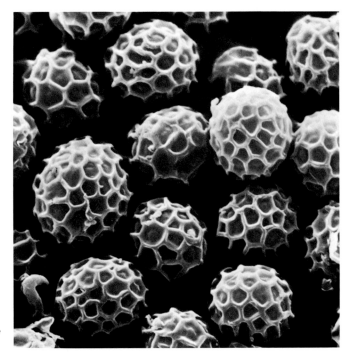

Reticulate chlamydospores with meshes of various sizes, mostly polygonal-shaped. SEM × 1500

21   *Ustilago avenae* (Persoon) Rostrup

Eumycota · Basidiomycotina · Teliomycetes · Ustilaginales · Ustilaginaceae

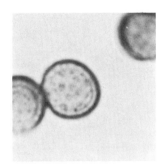

Subspherical chlamydospores in high focus. LM × 2000

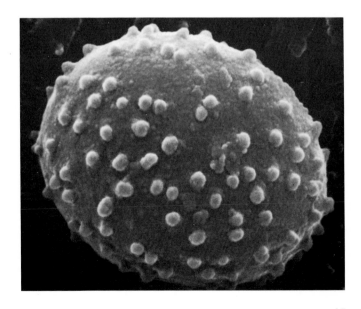

A broadly elliptic chlamydospore showing a finely
tuberculose surface with scattered hemispherical warts.
SEM × 12,000

22 *Ustilago maydis* (DC) Corda

Eumycota · Basidiomycotina · Teliomycetes · Ustilaginales · Ustilaginaceae

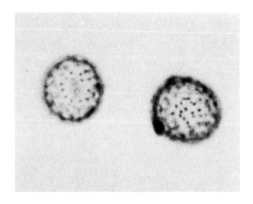

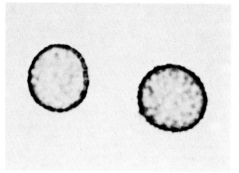

Chlamydospores in different foci. The upper two spores in high focus, the other in optical cross-section. LM × 2000

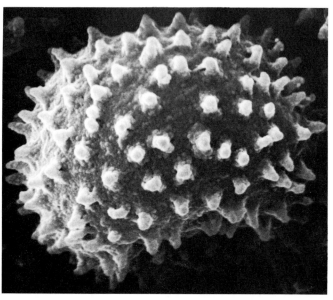

A chlamydospore with tuberculose surface bristling with subconical blunt spines. SEM × 10,000

Detail of irregularly nodulose spines and tuberculose spore surface. SEM × 25,000

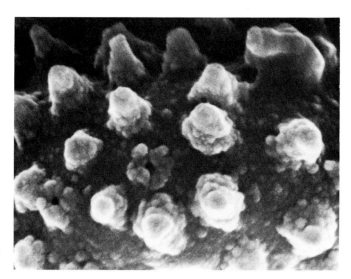

46

23 *Ustilago tragopogonis – pratensis* (Persoon) Roussel

Eumycota · Basidiomycotina · Teliomycetes · Ustilaginales · Ustilaginaceae

Spherical and reticulate chlamydospores in different foci: the chlamydospore to the left in high focus, the other one in optical cross-section. LM × 1900

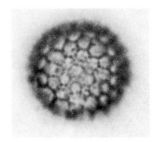 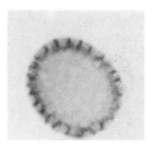

A chlamydospore with distinct reticulum of almost rounded lumina studded with irregular tubercles. SEM × 5300

Detail of the reticulum showing lumina studded with irregular tubercles. SEM × 8600

24 *Agaricus bitorquis* (Quélet) Saccardo

Eumycota · Basidiomycotina · Hymenomycetes · Agaricales · Agaricaceae

An ellipsoid basidiospore in sub-lateral view. LM ×2000

A basidiospore with smooth wall and a trapezoid proximal hilar appendix.
SEM ×12,000

25 *Amanita fulva* Schaeffer ex Persoon

Eumycota · Basidiomycotina · Hymenomycetes · Agaricales · Amanitaceae

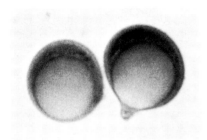

Two spherical basidiospores with guttula; the basidiospore on the right seen in adaxial-abaxial profile showing a prominent hilar appendix. LM × 2000

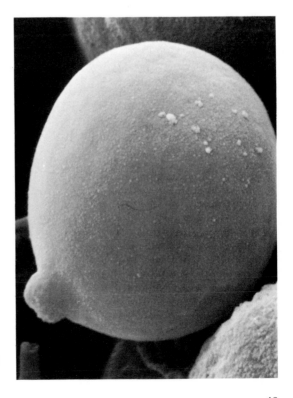

A spherical basidiospore in sublateral view, showing a short hilar appendix and a puncticulate wall. SEM × 8500

26 *Boletus edulis* Bulliard ex Fries

Eumycota · Basidiomycotina · Hymenomycetes · Agaricales · Boletaceae

Two basidiospores seen in adaxial-abaxial profile. LM × 2000

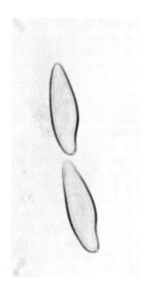

Basidiospores in lateral view showing the proximal rounded little hilar appendix and a puncticulate wall. SEM × 6000

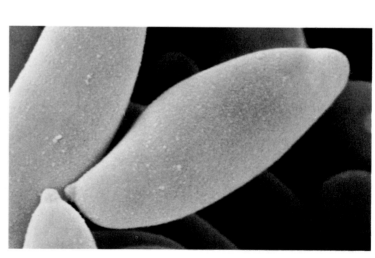

27 *Suillus grevillei* (Klotzsch) Singer

Eumycota · Basidiomycotina · Hymenomycetes · Agaricales · Boletaceae

A subfusoid basidiospore in adaxial-abaxial profile. LM ×2000

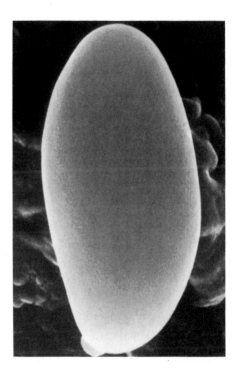

A fusiform basidiospore with a distinct hilar appendix and smooth wall. SEM ×9000

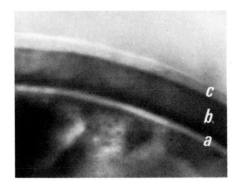

Section of the basidiospore wall with endosporium (*a*) electron-dense episporium and exosporium (*b*) and an electron-clear ectosporium (*c*). TEM ×20,000

28 *Cortinarius praestans* Cordier

Eumycota · Basidiomycotina · Hymenomycetes · Agaricales · Cortinariaceae

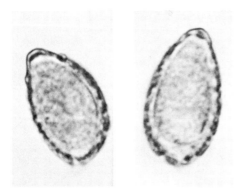

Irregularly verrucose basidiospore with distinct proximal hilar appendix and depressed hilar zone. SEM ×4000

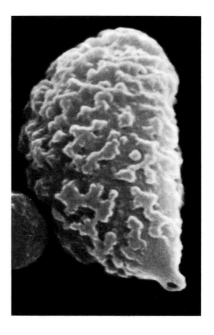

Two amygdaliform-fusoid basidiospores with verrucose ornamentation; in sublateral profile (*to the left*), by abaxial face (*to the right*). LM ×2000

Detail of verrucose ornamentation in proximal basidiospore part. SEM ×9000

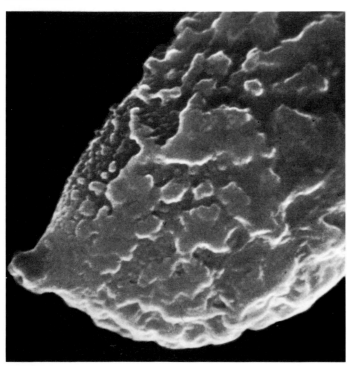

Section of basidiospore wall showing endosporium (*a*), episporium (*b*), irregularly thickened exosporium (*c*) with remnants of perisporium (*p*) and externally thin ectosporium (*d*), all around the multiguttulate cytoplasm. TEM ×20,000

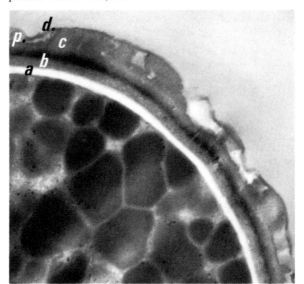

29 *Inocybe mixtilis* (Britzelmayr) Saccardo

Eumycota · Basidiomycotina · Hymenomycetes · Agaricales · Cortinariaceae

Gibbous basidiospores with subconical protuberances seen by
different views. LM × 2000

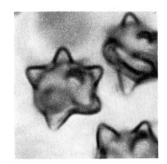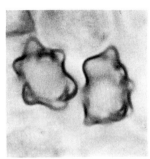

Two ellipsoid-nodulose basidiospores with a smooth
wall and little hilar appendix, in adaxial-abaxial profile. SEM × 7000

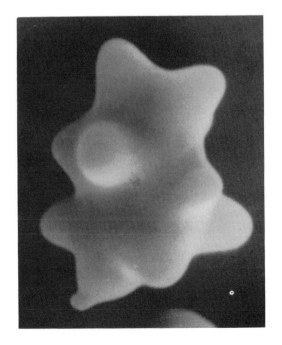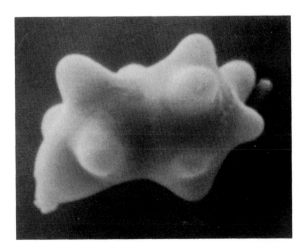

30 *Inocybe queletii* Maire & Konrad

Eumycota · Basidiomycotina · Hymenomycetes · Agaricales · Cortinariaceae

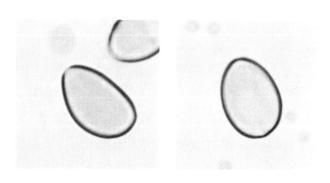

Basidiospores in lateral and frontal views; apical differentiation is seen in the smooth wall (*to the left*). LM × 2000

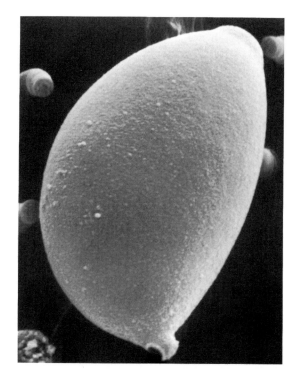

Minutely verrucose basidiospore in adaxial-abaxial profile, with lateral proximal hilar appendix where the hilar zone is depressed and with a distal, rounded differentiation. SEM × 8000

54

31 *Inocybe subcarpta* Boursier & Kühner

Eumycota · Basidiomycotina · Hymenomycetes · Agaricales · Cortinariaceae

A polygonal-gibbous basidiospore with large bosses. LM ×2000

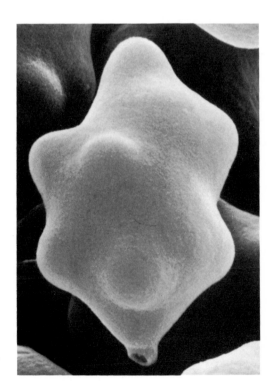

A basidiospore showing largely subconical, rounded protuberances, a short hilar appendix and a slightly scabrous surface. SEM ×8000

32 *Rozites caperata* (Persoon ex Fries) Karsten

Eumycota · Basidiomycotina · Hymenomycetes · Agaricales · Cortinariaceae

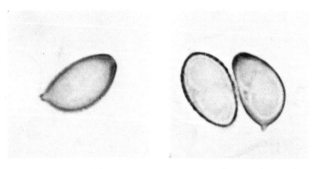

Basidiospores with verrucose ornamentation and tapering differentiated apex, seen in lateral and abaxial views. LM × 2000

A basidiospore showing a verrucose exosporial ornamentation with a nearly smooth papillate distal part. SEM × 7500

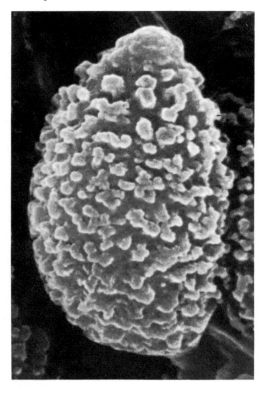

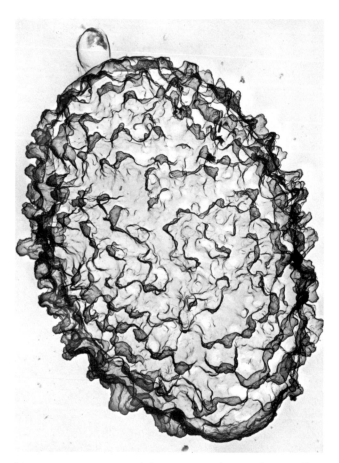

Verrucose ecto-exosporial ornamentation, carbon replica. TEM × 9500

Section showing a smooth fibrillous episporium (*b*), an irregularly verrucose exosporium (*c*) in a perisporium (*p*) and a thin ectosporial pellicle (*d*). TEM × 12,000

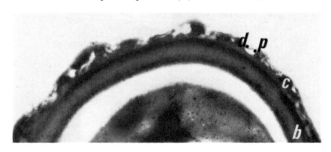

33 *Chroogomphus rutilus* (Schaeffer ex Fries) O. K. Miller

Eumycota · Basidiomycotina · Hymenomycetes · Agaricales · Gomphidiaceae

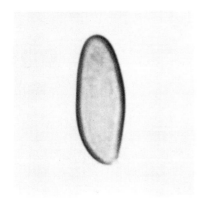

An elongate subfusiform basidiospore with proximal hilar appendix in lateral view. LM ×2000

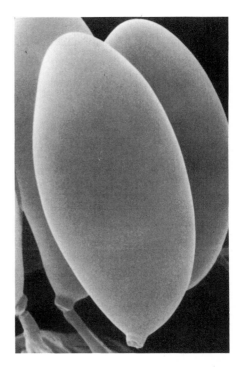

Basidiospores with smooth wall and a cylindrical hilar appendix. SEM ×5000

34 Clitopilus prunulus (Scopoli ex Fries) Kummer

Eumycota · Basidiomycotina · Hymenomycetes · Agaricales · Rhodophyllaceae

Ellipsoid-fusoid, somewhat angular basidiospores seen in lateral profile. LM ×2000

Basidiospore in sublateral view with prominent, longitudinal, meridional ridges. SEM ×12.000

Basidiospore in polar view, showing meridional ridges. SEM ×8000

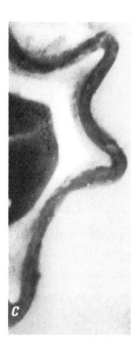

Section of basidiospore wall with smooth but angular electron-dark exosporium (*c*). TEM ×15,000

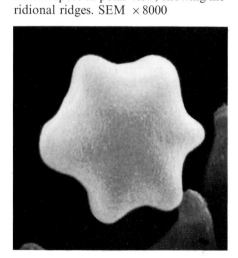

35 *Rhodophyllus hirtipes* (Schum. ex Fries) Lange

Eumycota · Basidiomycotina · Hymenomycetes · Agaricales · Rhodophyllaceae

A polyedral basidiospore seen in adaxial-abaxial profile. LM × 2000

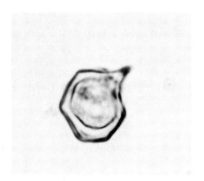

Polyedral basidiospores showing slightly depressed facets and a little truncate proximal hilar appendix. SEM × 5000 (*to the left*), × 9000 (*to the right*)

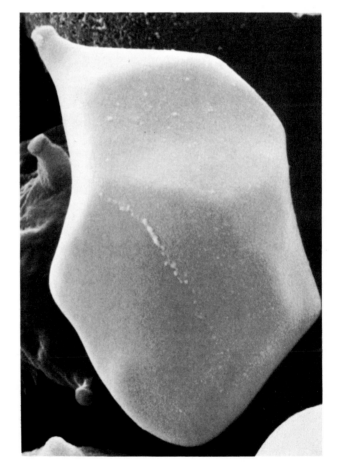

36 *Pluteus atricapillus* (Secr.) Sing.

Eumycota · Basidiomycotina · Hymenomycetes · Agaricales · Pluteaceae

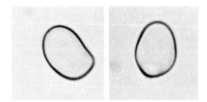 Two ellipsoid basidiospores with smooth wall: in sublateral profile (*to the left*), by adaxial face (*to the right*). LM ×2000

Smooth, ellipsoid basidiospore with a short hilar appendix. SEM ×13,000

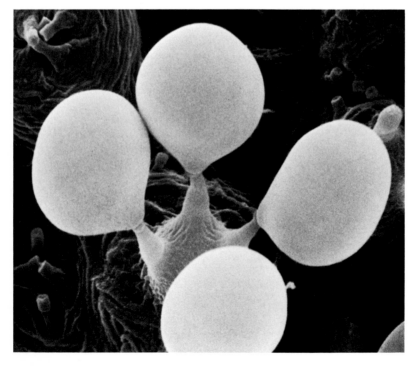

Four submature basidiospores on sterigmata. SEM ×3200

37 *Lactarius lignyotus* Fries

Eumycota · Basidiomycotina · Hymenomycetes · Agaricales · Russulaceae

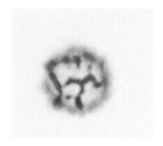 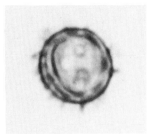

Two globose basidiospores in different foci: in high focus (*to the left*), in equatorial view (*to the right*). LM ×2000

Mature basidiospore with typical ornamentation consisting of irregular warts and crests sometimes forming an incomplete reticulum. SEM ×6000

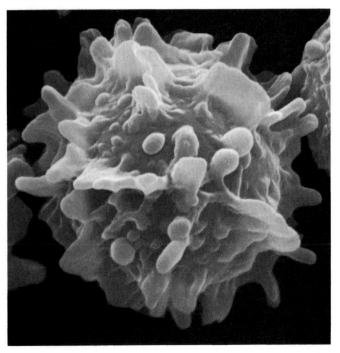

Three basidiospores on sterigmata, showing a distinct rounded-subconical, proximal hilar appendix and a smooth supra-appendicular area. SEM ×2000

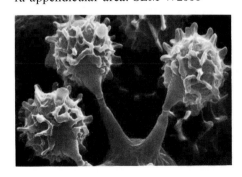

38 *Lactarius pterosporus* Romagnesi

Eumycota · Basidiomycotina · Hymenomycetes · Agaricales · Russulaceae

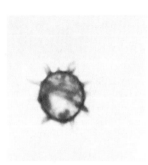

A subglobose basidiospore with prominent wing-like ornamentation. LM × 2000

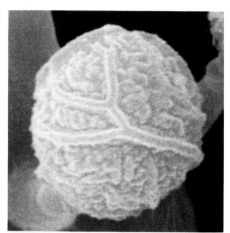

Principal stages in the development of the basidiospore orna- ▷ mentation during the maturation. SEM × 7000

A mature basidiospore showing an ornamentation of warts and columns between the anastomosed wing-like high crests. SEM × 8500

39 *Lactarius trivialis* Fries

Eumycota · Basidiomycotina · Hymenomycetes · Agaricales · Russulaceae

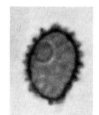

Warty and interrupto-reticulated basidiospores seen in sub-lateral view: in high focus (to the left), in plane of symmetry (to the right). LM ×2000

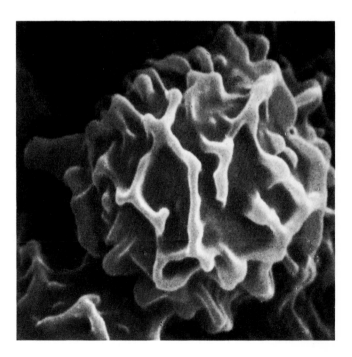

A broadly ellipsoid basidiospore with an ornamentation of subconical warts and festooned crests incompletely anastomosed. SEM ×10,000

40 *Russula emetica* (Schaeffer ex Fries) Persoon ex Fries

Eumycota · Basidiomycotina · Hymenomycetes · Agaricales · Russulaceae

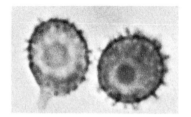

Warty basidiospores with guttula and elongated hilar appendix. LM ×2000

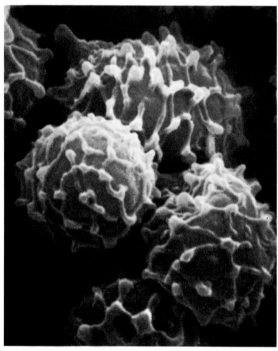

Basidiospores with typical ornamentation of subconical warts irregularly connected by lower crests. SEM ×5500

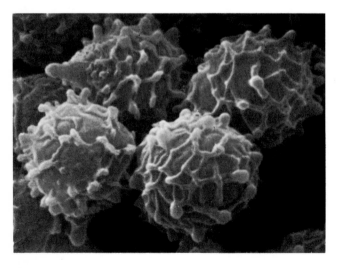

Group of basidiospores showing different aspects of interrupted reticulum. SEM ×4000

41 *Russula laurocerasi* Melzer var. *fragrans* Romagnesi

Eumycota · Basidiomycotina · Hymenomycetes · Agaricales · Russulaceae

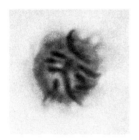 

Two highly ornamented basidiospores with sub-cylindrical hilar appendix. LM ×2000

A basidiospore in subfrontal view, showing an ornamentation consisting of distinct warts, short ridges and a rounded-subconical hilar appendix. SEM ×10,000

Section of the basidiospore wall, showing a smooth episporium (*b*) and a perisporial-ectosporial (*d*) veil covering exosporial ornaments. TEM ×7500

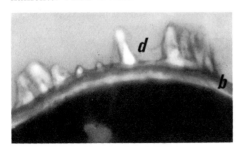

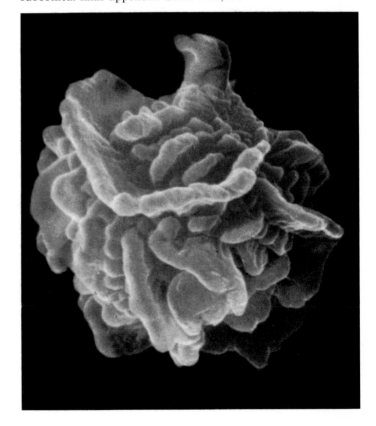

41

Russula laurocerasi Melzer var. *fragrans* Romagnesi

Eumycota · Basidiomycotina · Hymenomycetes · Agaricales · Russulaceae

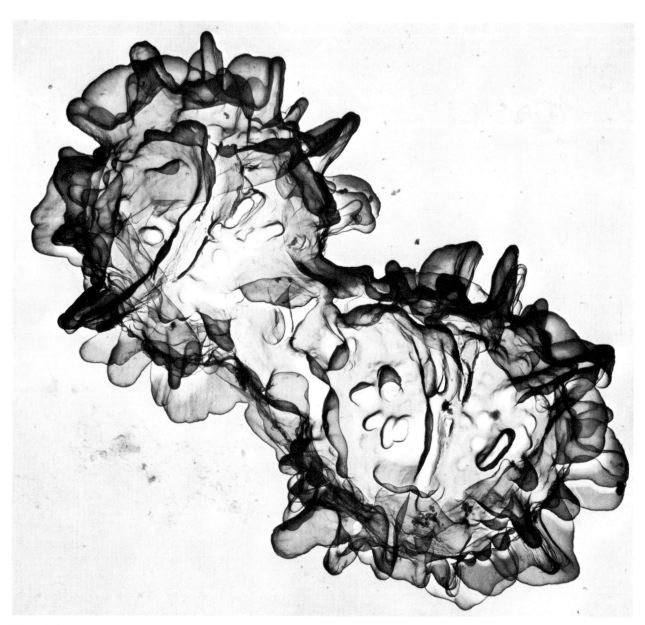

Two globose basidiospores with high ornamentation. TEM, carbon replica, ×10,000

42 *Russula puellaris* Fries

Eumycota · Basidiomycotina · Hymenomycetes · Agaricales · Russulaceae

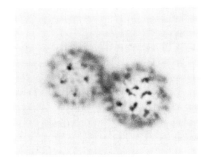 

Globose and subglobose basidiospores in different foci: in high focus (*to the left*), in equatorial view (*to the right*).

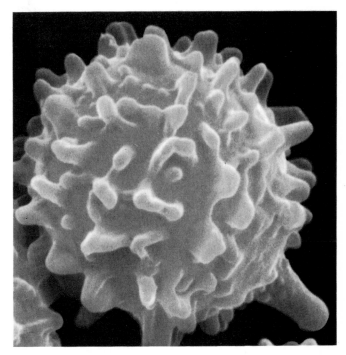

A basidiospore typically ornamented with warts and crests among some festooned wing-like ridges. SEM × 10,000

43   *Strobilomyces floccopus* (Vahl in Fl. Dan. ex Fries) Karsten

Eumycota · Basidiomycotina · Hymenomycetes · Agaricales · Strobilomycetaceae

Basidiospores in different foci: in high focus (*to the left*), in sublateral view (*to the right*).

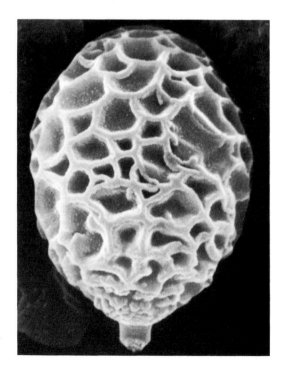

A mature ovoid basidiospore in frontal view, showing irregular and high network. SEM × 7000

44 *Naematoloma sublateritium* (Fries) Karsten

Eumycota · Basidiomycotina · Hymenomycetes · Agaricales · Strophariaceae

An ellipsoid and smooth basidiospore with a distal germ-pore, seen in lateral view. LM × 2000

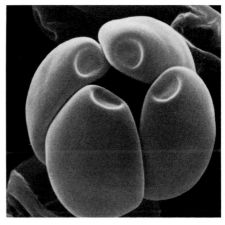

Four basidiospores in polar view with circular depressions marking the germ-pores. SEM × 7000

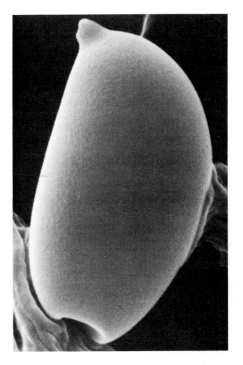

Basidiospore in adaxial-abaxial profile, showing a distinct hilar appendix and distal germ-pore. SEM × 10,000

45 *Lepista sordida* (Fries) Singer

Eumycota · Basidiomycotina · Hymenomycetes · Agaricales · Tricholomataceae

Two ellipsoid pustulate basidiospores in lateral view. LM ×2000

Group of sub-mature ellipsoid basidiospores irregularly ornamented with pustules of various sizes. SEM ×12,000

46 *Melanoleuca evenosa* (Saccardo) Konrad

Eumycota · Basidiomycotina · Hymenomycetes · Agaricales · Tricholomataceae

Oblong-ellipsoid, verrucose basidiospore in lateral view. LM ×2000

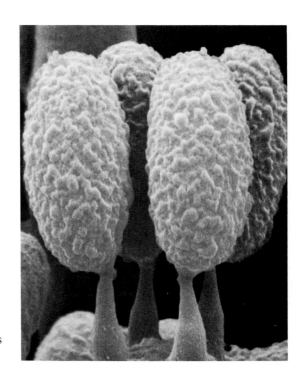

Four sub-mature basidiospores on sterigmata, showing numerous irregular rounded and more or less confluent warts. SEM ×6000

47 *Oudemansiella radicata* (Rehlan ex Fries) Singer

Eumycota · Basidiomycotina · Hymenomycetes · Agaricales · Tricholomataceae

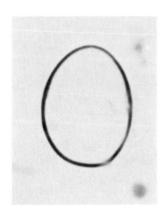

Ellipsoid basidiospore in sublateral profile. LM ×2000

Basidiospore on sterigma, showing a minutely warted, almost vermiculate surface. SEM ×5000

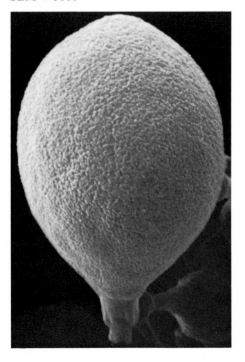

Basidiospore ornamentation consisting of numerous irregular rounded tubercles. SEM ×5000

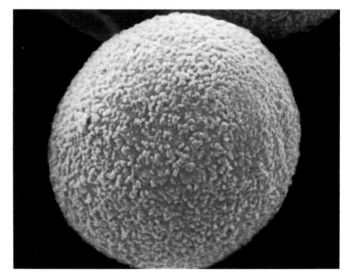

48 *Tricholomopsis platyphylla* (Persoon ex Fries) Singer

Eumycota · Basidiomycotina · Hymenomycetes · Agaricales · Tricholomataceae

Broadly ellipsoid smooth basidiospore seen by adaxial-abaxial profile. LM ×2000

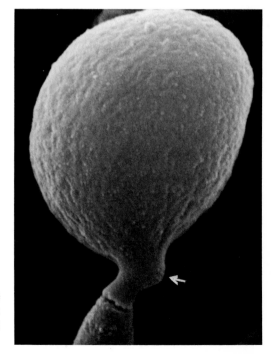

Basidiospore on sterigma just before detachement; the foveate surface seems to be an artefact. At the proximal part, there is a hilar appendix with "*punctum lacrymans*" (*arrow* on the apophyse) and pedicel with hilar zone. SEM ×4500

49 *Xeromphalina campanella* (Batsch ex Fries) Kühner & Maire

Eumycota · Basidiomycotina · Hymenomycetes · Agaricales · Tricholomataceae

Ellipsoid – sub-cylindrical basidiospore in almost lateral view. LM × 2000

Basidiospores on sterigmata showing a rugulose-plicate surface and distinct proximal hilar appendix. SEM × 8000

50 *Cantharellus cibarius* Fries

Eumycota · Basidiomycotina · Hymenomycetes · Aphyllophorales · Cantharellaceae

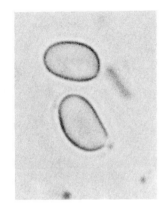

Ellipsoid, smooth basidiospores seen in adaxial-abaxial profile. LM × 2000

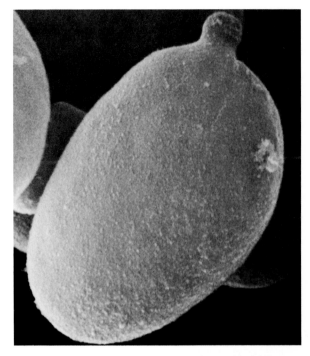

An almost smooth basidiospore with rounded hilar appendix. SEM × 10,000

Section of the basidiospore wall showing a granulose outer ectosporium (*d*), an electron-clear perisporium (*p*) and an episporium (*b*) around the cytoplasm. TEM × 50,000

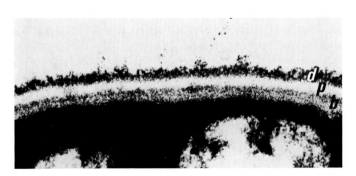

51 Serpula lacrymans (Wulfen ex Fries) Karsten

Eumycota · Basidiomycotina · Hymenomycetes · Aphyllophorales · Coniophoraceae

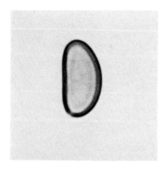

A sub-ellipsoid basidiospore with a slightly depressed adaxial face and a distinct wall, seen in adaxial-abaxial profile. LM ×2000

Section of the basidiospore wall showing two layers, episporium (*b*) and exosporium (*c*) beneath a granulose ectosporium. TEM ×80,000

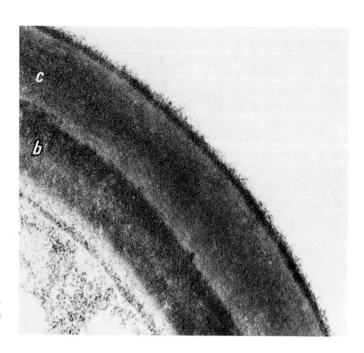

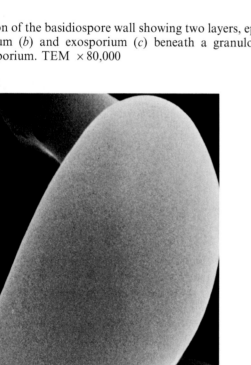

Two basidiospores with a smooth surface, the one to the right with a prominent hilar appendix. SEM ×9000

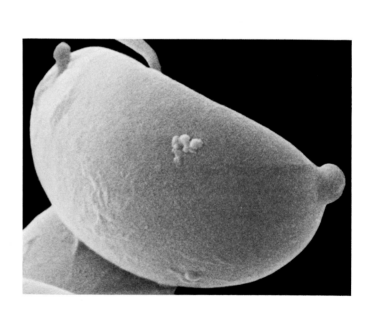

52 *Peniophora quercina* Fries

Eumycota · Basidiomycotina · Hymenomycetes · Aphyllophorales · Corticiaceae

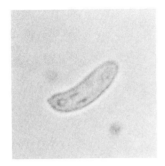

An allantoid, smooth basidiospore seen in adaxial-abaxial profile. LM × 2000

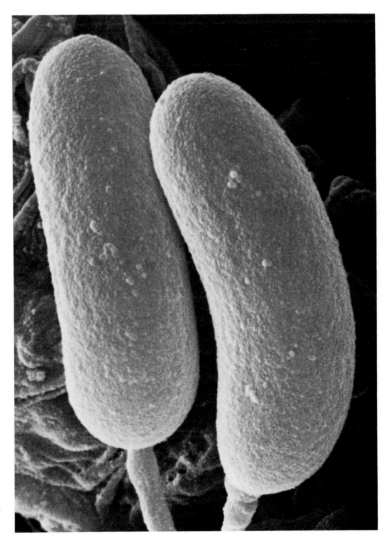

Basidiospores on sterigmata, showing a minutely irregular warted surface. SEM × 10,000

53 Ganoderma applanatum (Persoon ex S. F. Gray) Patouillard

Eumycota · Basidiomycotina · Hymenomycetes · Aphyllophorales · Ganodermataceae

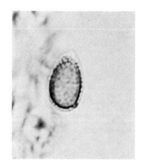

An ellipsoid and foveate basidio-spore seen in adaxial-abaxial profile. LM ×2000

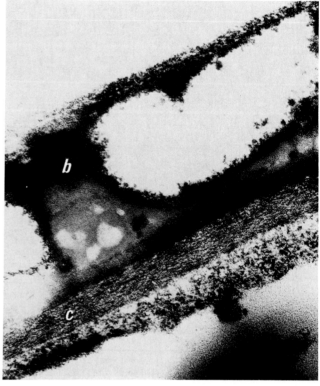

Section through the basidiospore wall showing, cytoplasm, an episporium (*c*), a thick exosporium (*b*) with pillars and cavities. TEM ×80,000

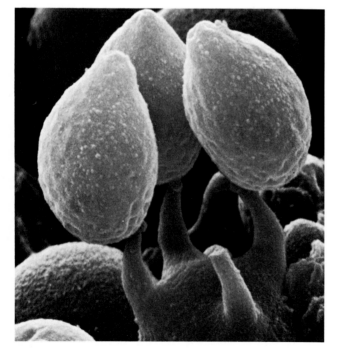

Three submature basidiospores on sterigmata showing a foveate ornamentation, a proximal rounded hilar appendix and an elongate, truncated apical part. SEM ×8000

54 *Gomphus clavatus* (Persoon ex Fries) S. F. Gray

Eumycota · Basidiomycotina · Hymenomycetes · Aphyllophorales · Gomphaceae

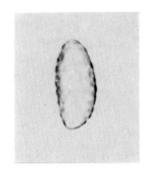

A pustulate sub-fusiform basidiospore seen in adaxial-abaxial profile. LM ×2000

An irregularly pustulate basidiospore showing a large, non-delimitated hilar appendix. SEM ×10,000

Section of basidiospore wall showing externally a pustulate exosporium (*c*), lenticular remnants of a perisporium (*p*), an episporium (*b*) and an endosporium (*a*) above the multiguttulate cytoplasm. TEM ×30,000

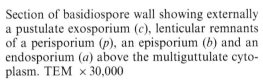

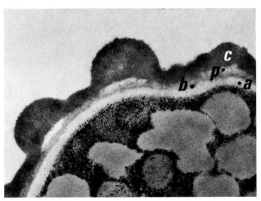

55 *Hydnum repandum* L. ex Fries

Eumycota · Basidiomycotina · Hymenomycetes · Aphyllophorales · Hydnaceae

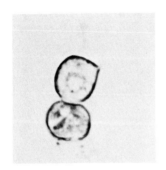

Two spherical basidiospores with distinct hilar appendix. LM ×2000

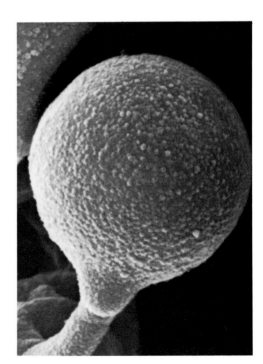

A sub-mature basidiospore still bound to the sterigma and showing a surface irregularly ornamented with minute, rounded warts. SEM ×8000

56 *Heterobasidion annosum* (Fries) Brefeld

Eumycota · Basidiomycotina · Hymenomycetes · Aphyllophorales · Polyporaceae

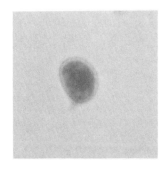 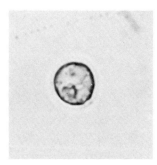

Subglobose basidiospores seen in adaxial-abaxial profile and polar view. LM ×2000

A subglobose basidiospore ornamented with irregular tubercles (adaxial face) and showing a distinct proximal hilar appendix. SEM ×17,000

Section of basidiospore wall showing an ornamented electron-dark exosporium (*c*), a thin perisporial layer (*p*) and a fibrillous episporium (*b*). TEM ×35,000

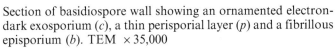

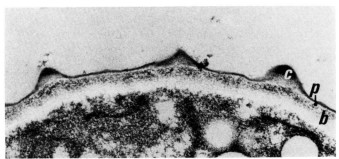

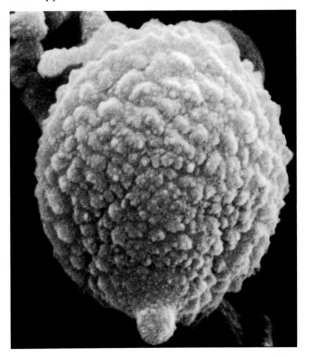

81

57 *Polyporus squamosus* (Hudson ex Fries) Fries

Eumycota · Basidiomycotina · Hymenomycetes · Aphyllophorales · Polyporaceae

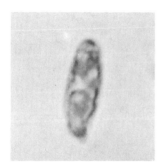

An elongate-ellipsoid smooth basidiospore seen in adaxial-abaxial profile. LM ×2000

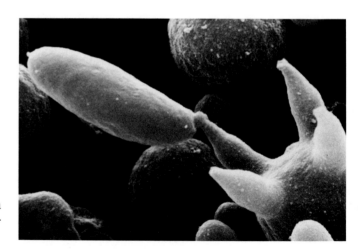

A sub-mature basidiospore on the sterigma showing a smooth wall and a distinct rounded, proximal hilar appendix. SEM ×4000

58 *Phylacteria terrestris* (Ehrhart ex Fries) Patouillard

Eumycota · Basidiomycotina · Hymenomycetes · Aphyllophorales · Thelephoraceae

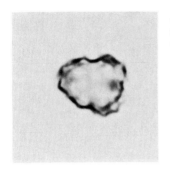

A spinulose-gibbous basidiospore in equatorial view. LM ×2000

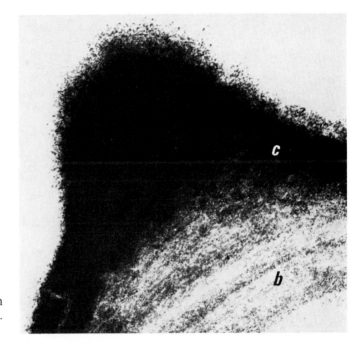

Section through a spine with a wall composed of an electron-dark exosporium (*c*) and an episporium (*b*). TEM ×100,000

A gibbous basidiospore with groups of spines and an irregularly warted surface. SEM ×8000

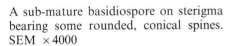

A sub-mature basidiospore on sterigma bearing some rounded, conical spines. SEM ×4000

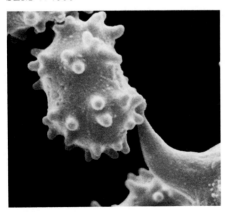

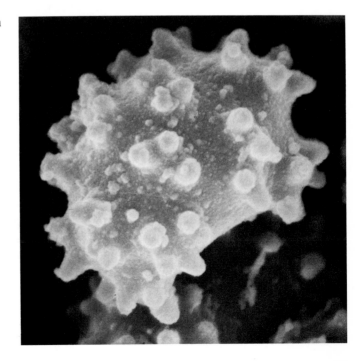

59 *Bovista plumbea* Persoon per Persoon

Eumycota · Basidiomycotina · Gasteromycetes · Lycoperdales · Lycoperdaceae

A subglobose basidiospore with long pedicel. LM × 2000

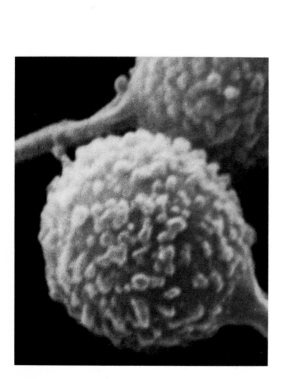

Two minutely and irregularly warty basidiospores with pedicel. SEM × 6000

Detail of the irregular and verrucose ornamentation. SEM × 10,000

60 *Calvatia excipuliformis* (Scopoli trans Persoon cum emend.) Perdeck

Eumycota · Basidiomycotina · Gasteromycetes · Lycoperdales · Lycoperdaceae

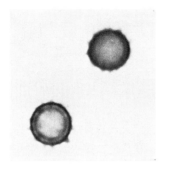

Two spherical, verrucose basidiospores in equatorial view. LM ×2000

Detail of the ornamentation: slightly granulate surface and columns with tabular top. SEM ×20,000

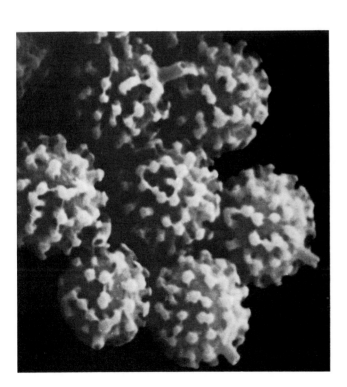

Group of basidiospores showing an ornamentation of small flattened columns and cylindrical pedicels. SEM ×5000

61 *Lycoperdon perlatum* Persoon per Persoon

Eumycota · Basidiomycotina · Gasteromycetes · Lycoperdales · Lycoperdaceae

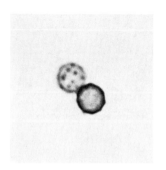

Two globose and verrucose basidiospores in high focus and equatorial view. LM ×2000

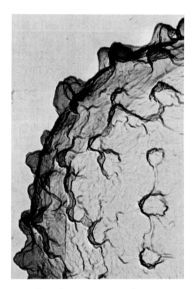

Detail of ornamentation: exosporial warts are covered by a perisporial-ectosporial veil.
TEM, carbon replica, ×20,000

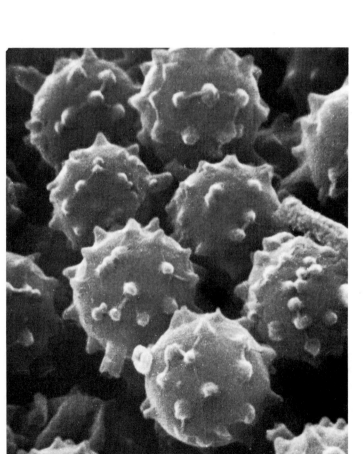

Group of basidiospores with ornamentation consisting of isolated or connected subconical spines.
SEM ×8500

62 *Acremonium butyri* (van Beyma) Gams

Eumycota · Deuteromycotina

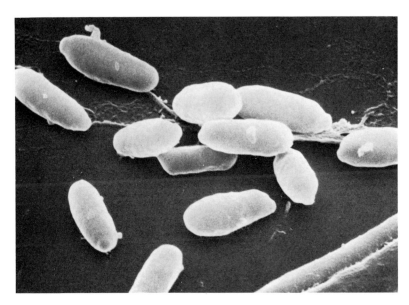

Smooth, elongate conidia with rounded ends and slightly prominent hilum. SEM ×7500 (*above*), ×18,500 (*below*)

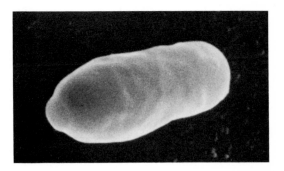

63 *Alternaria* sp.

Eumycota · Deuteromycotina

Conidia. LM ×1000

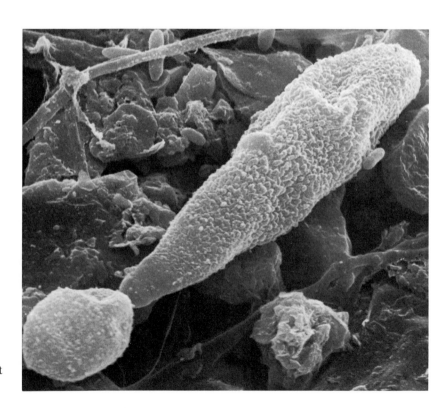

Mature conidium with young conidium at
the apex. SEM ×3100

64 *Arthrinium* sp.

Eumycota · Deuteromycotina

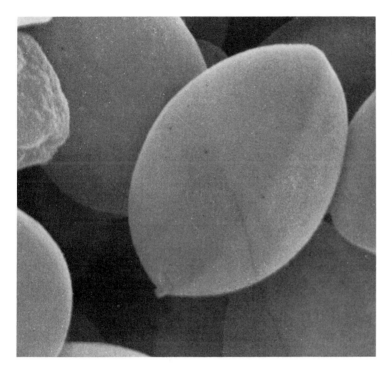

Radulaspores (basauxic), smooth wall with longitudinal ring. Hilum is visible at the top as a short denticle. SEM ×12,300

65 *Aspergillus sp.*

Eumycota · Deuteromycotina

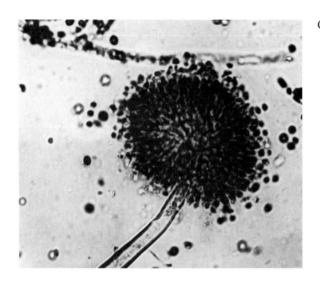

Conidial head. LM × 1200

Smooth conidia in chains. SEM × 8000

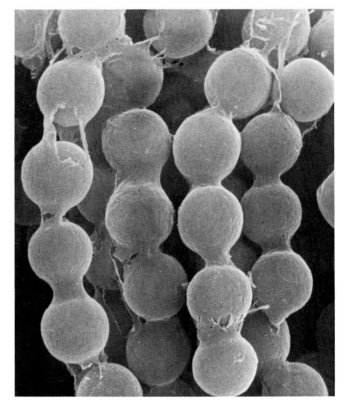

Ornamentation of conidia. SEM × 3100

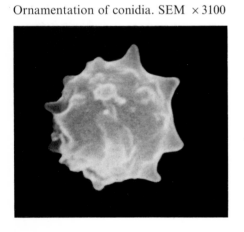

66 *Botrytis cinerea* Pers. ex Nocca & Balb.

Eumycota · Deuteromycotina

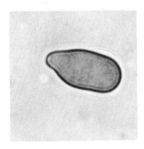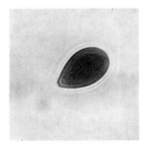

Conidia with basal scar. LM ×1300

Ornamentation of conidia.
SEM ×3100

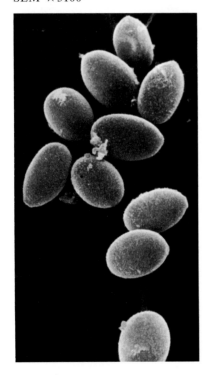

Conidia in higher magnification. SEM ×6500

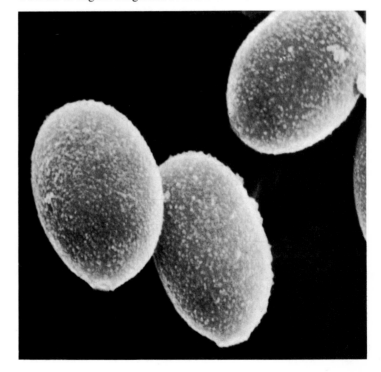

67 *Briosia cubispora* (Berk. & Curt.) v. Arx

Eumycota · Deuteromycotina

Chains of conidia. SEM × 2200

Conidia in higher magnification. SEM × 5000

68 *Chalara elegans* Nag Raj & Kendrick

Eumycota · Deuteromycotina

Septate aleuriospores with hyaline basal cells. LM ×800. (Courtesy of Institut du Tabac, Bergerac, France)

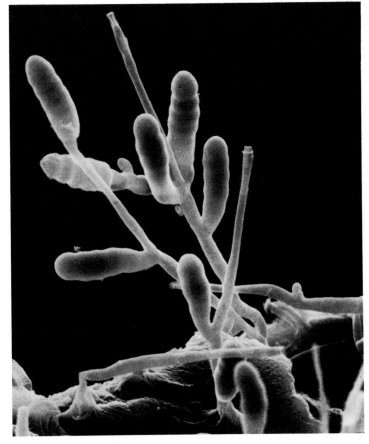

Phialides and aleuriospores, restricted at septa. SEM ×880

68 *Chalara elegans* Nag Raj & Kendrick

Eumycota · Deuteromycotina

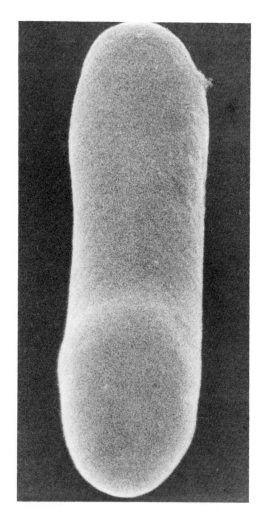

Detached phialospore. SEM ×11,400

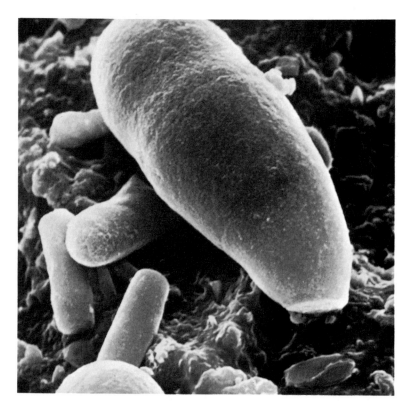

Detached aleuriospore, constricted at septa.
SEM ×4400

69 *Cladosporium cladosporioides* (Fres.) De Vries

Eumycota · Deuteromycotina

Apex of conidiophore with two ramoconidia and small fragments of conidial chains (*arrow*). LM ×1200

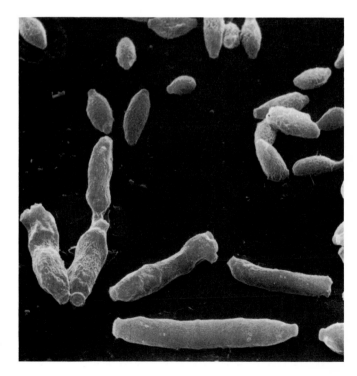

Separated or attached, slightly ornamented conidia. Note truncate scars. SEM ×2900

70 *Conoplea mangenotii* Reisinger

Eumycota · Deuteromycotina

Conidium on a conidiophore.
LM × 1750

Smooth conidia, hilum and invagination at the level of the germ slit. SEM × 2100

Smooth conidia. SEM × 12,000

70 *Conoplea mangenotii* Reisinger

Eumycota · Deuteromycotina

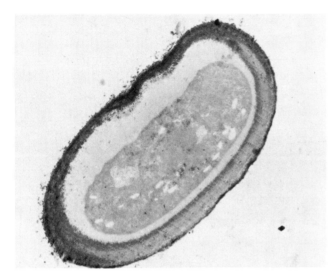 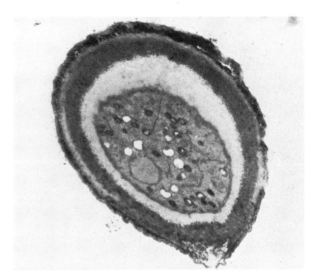

Sections of conidia showing wall and hilum structure. TEM × 9100 (*to the left*), × 8500 (*to the right*)

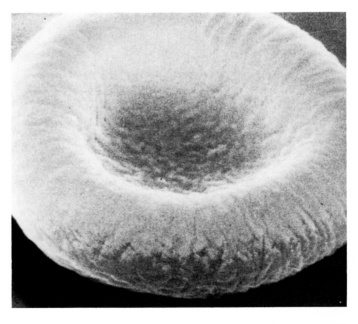

Smooth conidia. SEM × 12,000

97

71 *Deightoniella torulosa* (Syd.) M.B. Ellis

Eumycota · Deuteromycotina

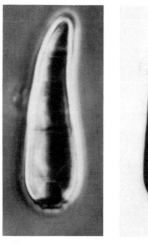 

Shape of conidia, anatomy and basal hilum. LM ×1000

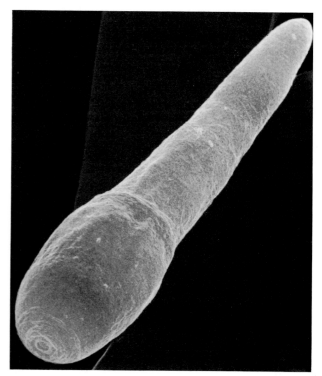

Conidium with smooth surface and hilum. SEM ×2200

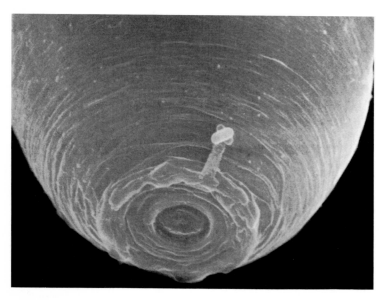

The same at higher magnification. SEM ×10,000

72 *Dendryphiella vinosa* (Berk. & Curt.) Reisinger

Eumycota · Deuteromycotina

Section of basal conidium. LM × 1900

Terminal conidium.
LM × 5200

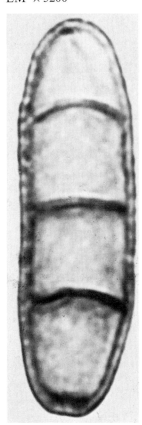

Wall ornamentation and scar of insertion point between two conidia. SEM × 11,600

Group of conidia. SEM × 5000

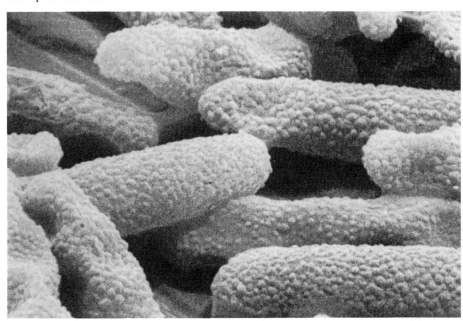

72 *Dendryphiella vinosa* (Berk. & Curt.) Reisinger

Eumycota · Deuteromycotina

Conidial wall structure. TEM × 4500 (*to the left*), × 44,000 (*to the right*)

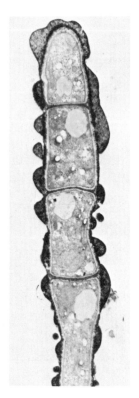

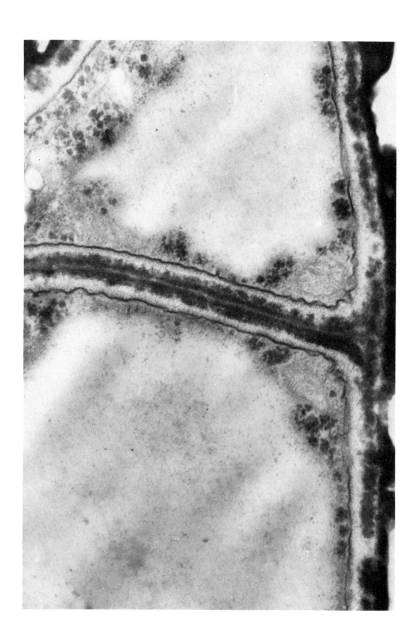

73 *Drechslera avenae* (Eidam) Sharif.

Eumycota · Deuteromycotina

Phragmoconidia (4–6 cross walls), distoseptate. Wall more or less melanized according to age. Thinner zone of decreased resistance (*arrow*). LM × 860

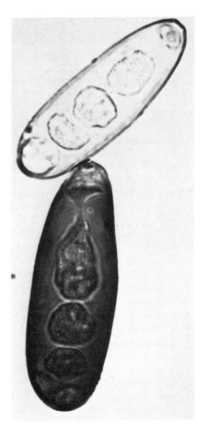

Phragmoconidia. LM × 1100

73 *Drechslera avenae* (Eidam) Sharif.

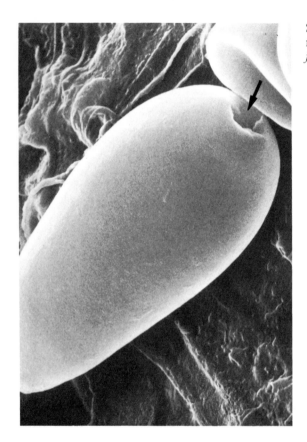

Shape of conidia with smooth surface. Apical, less resistant invaginated zone (*arrow*, *upper figure*, SEM ×2500) hilum (*arrow*, *lower figure*, SEM ×2500)

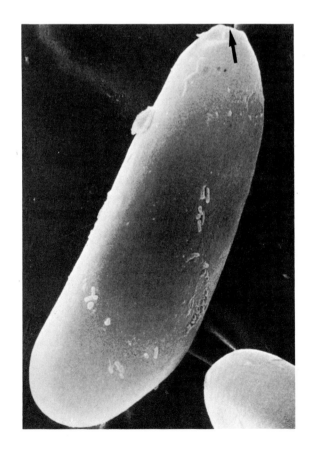

74 *Drechslera spicifera* (Bain.) Nicot

Eumycota · Deuteromycotina

Conidium. LM × 1500

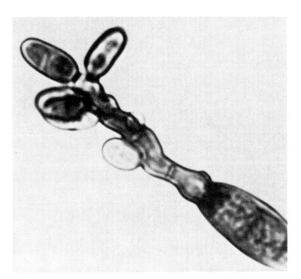

Iterative germination. LM × 800

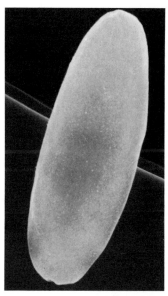

Smooth-walled conidium.
SEM × 1700

Desiccated conidium with retracted hilum. SEM × 11,000

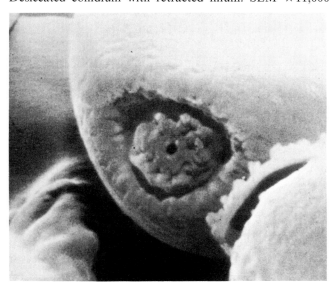

Longitudinal section of conidium wall and septum.
TEM × 16,000

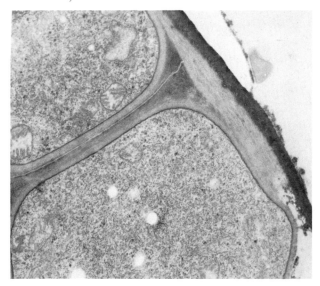

75 *Fusarium oxysporum* Schlecht. ex Fr. f. sp. *lycopersici*

Eumycota · Deuteromycotina

Shape and septation of conidia (unicellular conidia = microconidia). LM × 300

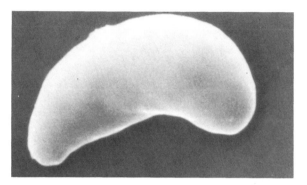

Unicellular microconidia with smooth, hyaline wall. SEM × 6500

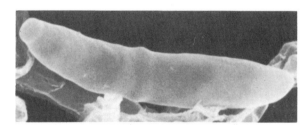

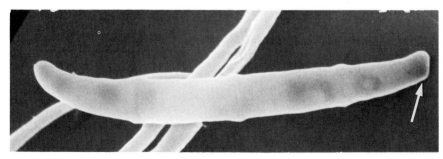

Markedly septate macroconidia. Foot-cell with a peculiar shape (*arrow*). SEM × 2900

76   *Geotrichum candidum* Link

Eumycota · Deuteromycotina

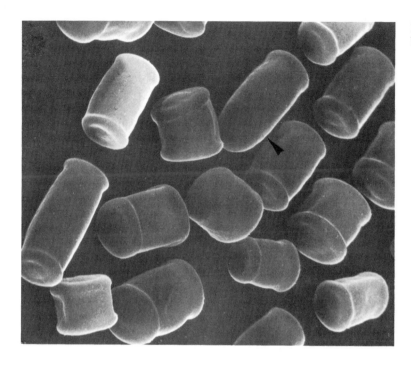

Free conidia; a terminal conidium characterized by a rounded apex (*arrow*). SEM × 5000

77 *Gliocladium catenulatum* Gilm. & Abbott

Eumycota · Deuteromycotina

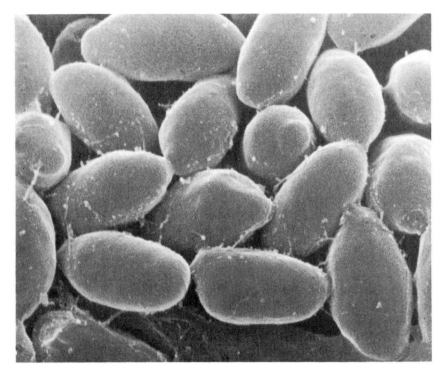

Conidia smooth, locally ornamented by mucous deposits, sometimes laterally flattened with large slightly prominent hilum. SEM ×9500

78 *Gliomastix murorum* (Corda) Hughes

Eumycota · Deuteromycotina

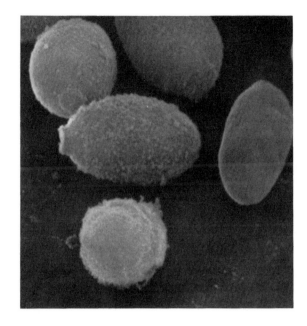

Unicellular, smooth to rough, ovoid conidia with prominent hilum.
SEM × 9900

79 *Metarrhizium anisopliae* (Metsch.) Sorok.

Eumycota · Deuteromycotina

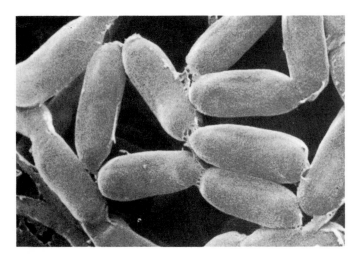

Unicellular, cylindrical, smooth or finely ornamented conidia showing lateral and terminal adherences probably due to surface mucus, small-spored strain. SEM × 2400

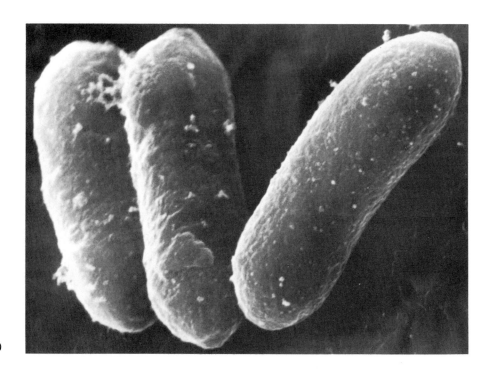

Conidia with a mucous surface,
large-spored strain. SEM × 5500

80 *Nigrospora oryzae* (Berk. & Br.) Petch

Eumycota · Deuteromycotina

Unicellular, smooth, pill-shaped conidium in equatorial view on a conidiophore. SEM ×4600

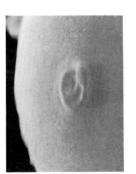

Conidial hilum. SEM ×10,000

Liberated conidia. SEM ×2000

81 *Paecilomyces elegans* (Corda) Mason & Hughes

Eumycota · Deuteromycotina

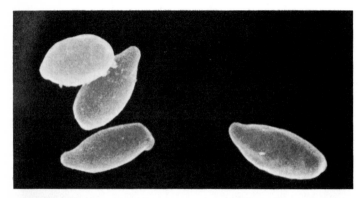

Unicellular, hyaline conidia showing truncate base and pointed apex. Smooth with ornamentations probably due to mucous deposits. SEM × 7000 (*above*), SEM × 16,000 (*below*)

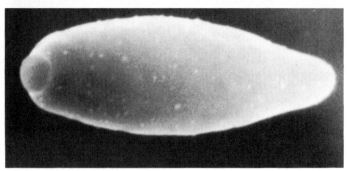

82 *Penicillium* sp.

Eumycota · Deuteromycotina

Conidia on sporulating structure.
LM × 1600

Conidia on the leg of *Tyrophagus putrescentiae*, a mite. SEM × 4500

Ornamented isolated conidia. SEM × 15,000

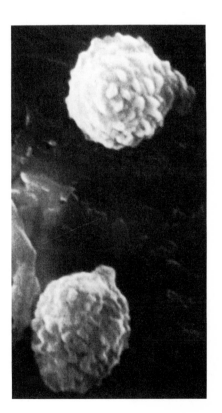

Ornamented conidia in chains. SEM × 12,000

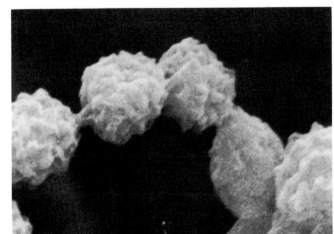

83 *Scopulariopsis brevicaulis* (Sacc.) Bainier

Eumycota · Deuteromycotina

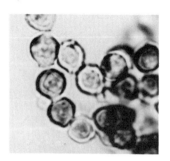

Conidial chains. LM × 1000

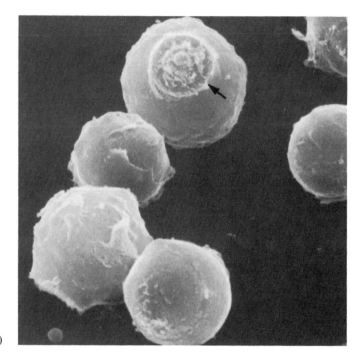

Conidia with flattened hilum (*arrow*). SEM × 3500

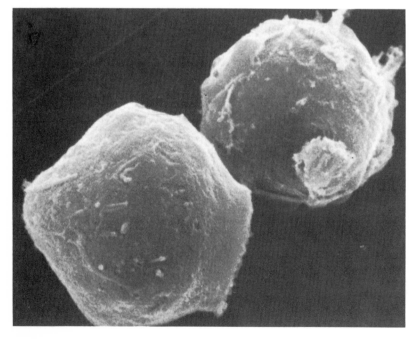

Two conidia. SEM × 5200

84 *Scytalidium lignicola* Pesante

Eumycota · Deuteromycotina

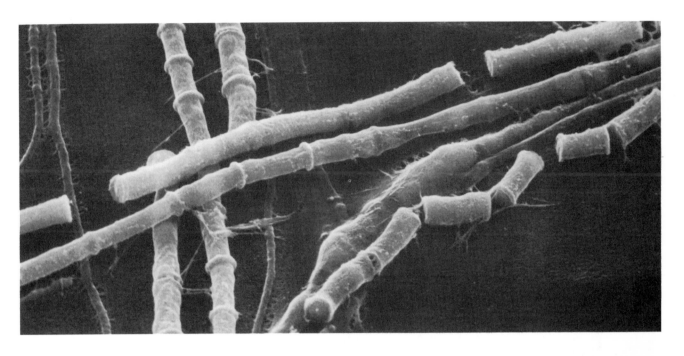

Transformation of vegetative hyphae into conidia; regularly spaced rims reveal the presence of cross walls. SEM × 500

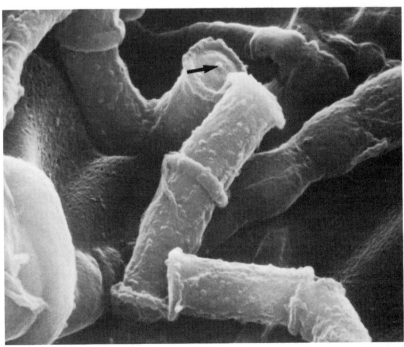

Some separated conidia with plugged cross-wall pore (*arrow*). SEM × 1000

85 *Stachybotrys chartarum* (Ehrenb. ex Link) Hughes

Eumycota · Deuteromycotina

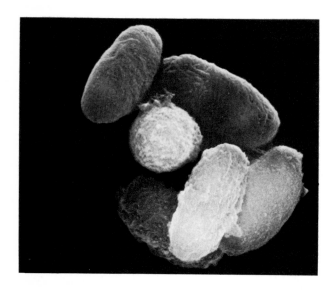

Unicellular, slightly ornamented conidia. Conidial head gathered by mucous substance. SEM ×4600

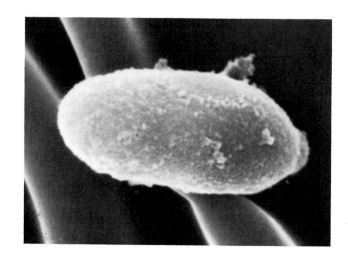

Conidium with ornamentation and prominent, truncated hilum. SEM ×13,000

86 *Trichocladium asperum* Harz

Eumycota · Deuteromycotina

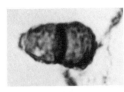

Shape and septation of conidia. LM × 720

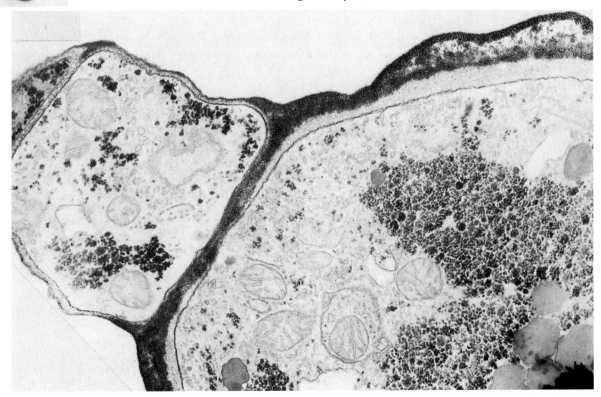

Section showing wall layers, ornamentation and hilum. TEM × 15,000

86 *Trichocladium asperum* Harz

Eumycota · Deuteromycotina

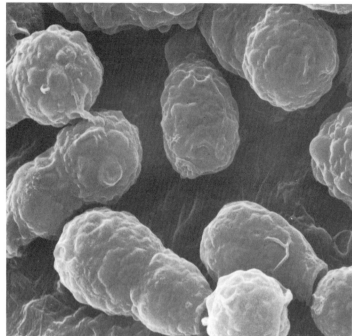

Conidia with warty ornamentation. SEM ×2000

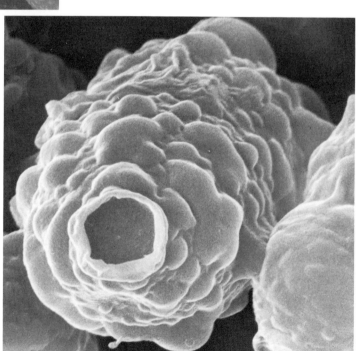

A conidium with warty ornamentation and a wide hilum. SEM ×6000

87 *Trichothecium roseum* (Pers.) Link ex S. F. Gray

Eumycota · Deuteromycotina

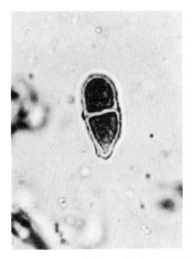

A septate conidium. LM × 1000

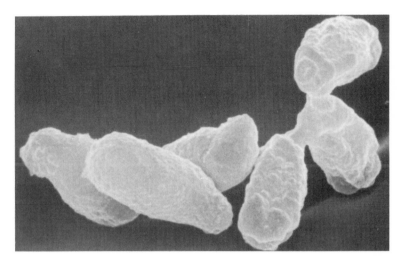

Chain of attached conidia. SEM × 1700

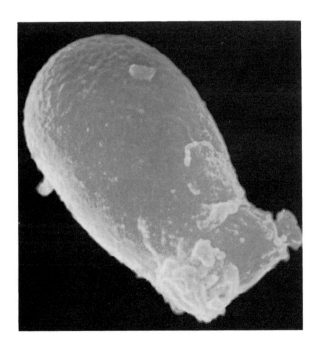

Conidium with two lateral flattened scars. SEM × 4500

Spore Descriptions

MYXOMYCOTA

    Liceales   *Lycogala epidendrum* (L.) Fries   1
      Reticulariaceae

Encysted spores pinkish grey in mass, changing to ochraceous or pallid with age, colourless by transmitted light; 6–7.5 μm in diameter; globose; wall composed of three layers: a thick and relatively electron-dense inner one, a thin and electron-clear middle one and a thin electron-dense outer one which bears the partially or completely reticulate ornamentation of irregularly shaped meshes.

    Trichiales   *Arcyria denudata* (L.) Wettst.   2
      Trichiaceae

Encysted spores red or reddish brown in mass, colourless by transmitted light; 6–8 μm in diameter; globose or subglobose; wall thin, with numerous irregular subhemispherical warts and a few tuberculose processes.

Trichia floriformis (Schw.) G. Lister   3

Encysted spores brick red in mass, pale red by transmitted light; 10–12 μm in diameter; ellipsoid or globose; wall thin, very densely covered with irregular rough processes.

Trichia scabra Rostrup   4

Encysted spores yellow or orange in mass, yellow by transmitted light; (9)–10–12 μm in diameter; globose; wall thin, marked by a delicately meshed and distorted reticulum.

EUMYCOTA

    Zygomycotina   *Mortierella ramanniana* (Moeller) Linnemann   5
      Mucorales
        Mortierellaceae

Sporangiospores colourless, but may seem pinkish-greyish in mass by the remains of sporangium walls, colourless in transmitted light, strongly refractive; 2–3 μm in diameter, minute; almost spherical to shortly elliptical; wall structure seems smooth by transmitted light; exospore with dense warts and spines in a net-like structure (SEM).

        Mucoraceae   *Rhizopus rhizopodiformis* (Cohn) Zopf   6

Sporangiospores colourless but may seem greyish-black in mass by the remains of sporangium walls, colourless in transmitted light, strongly refractive; (4.5) 5–6 (9) μm in diameter; shape and diameter

Mucoraceae

rather variable; mostly spherical to slightly oval to slightly poly-hedral; exosporium with spinules and ridges (SEM).

Rhizopus stolonifer (Ehrenb. ex Fr.) Lind   7

Sporangiospores colourless but may seem greyish to blackish in mass by the remains of sporangium walls, colourless in transmitted light, strongly refractive; 10–15 (20) × 7–9 (12) μm; shape and diameter rather variable, irregularly broadly elliptical to almost spherical; wall in dry spores with conspicuous exosporial longitudinal ridges, also seen by transmitted light; the ridges more or less disappear after swelling in water.

Ascomycotina
 Helotiales
  Helotiaceae

Bulgaria inquinans Pers. ex Fr.   8

Ascospores very dark brown in mass, in transmitted light very dark brown, opaque. In every ascus four dark brown and four similar, but hyaline ascospores are produced; 11–14 × 6–7 μm; spores somewhat kidney-shaped, non-septate; spore wall smooth, fibrous.

Pezizales
 Helvellaceae

Helvella crispa Fr.   9

Ascospores colourless in mass, in transmitted light hyaline; 16–20 × 10–13 μm; spores broadly elliptical, with one large central oil drop; spore wall smooth; exosporium fibrous, coarsely net-like (SEM).

Rhizina undulata Fr.   10

Ascospores slightly cream-coloured in mass, in transmitted light hyaline. 22–40 × 8–11 μm; ascospores fusiform, apiculate, with several oil drops; spore wall smooth, somewhat coarse (SEM).

Pezizaceae

Otidea onotica (Pers.) Fuckel   11

Ascospores white in mass, in transmitted light hyaline, 12–13 × 5–6 μm; ascospores broadly elliptical, with two large oil drops; spore wall smooth; exosporium rather smooth, with scattered low warts (SEM).

120

Plectascales   *Elaphomyces granulatus* Fr.   12
    Elaphomycetaceae

Ascospores black in mass, in transmitted light very dark brown to black, opaque; 24–32 µm in diameter; ascospores spherical, large; spore wall covered with irregular wart-like processes; exospore with teeth-like inflattened warts (SEM).

Sphaeriales   *Chaetomium globosum* Kunze ex Fr.   13
    Melanosporaceae

Ascospores dark olive brown in mass and in transmitted light; 11–13 × 8–10 µm; spores broadly lemon-shaped, apiculate at both ends; non-septate; wall structure net-like, with scattered low warts.

Nectriaceae   *Nectria cinnabarina* (Tode ex Fr.) Fr.   14

Ascospores colourless, hyaline in transmitted light; 12–15 × 4–9 µm; ascospores elliptic, slightly fusiform, with a median transverse septum; wall smooth in transmitted light, exoperidium with a rough, low-warty structure (SEM).
Conidia (St. conid. *Tubercularia vulgaris* Tode ex Fr.) colourless, hyaline in transmitted light; 6–8 × 1.5–2 µm; small, cylindrical, with several small vacuoles; non-septate; cell wall smooth.

Sphaeriaceae   *Ustulina deusta* (Fr.) Petrak   15

Ascospores very dark brown in mass, in transmitted light very dark brown, opaque; 28–34 × 7–10 µm; ascospores fusiform, with one side somewhat flattened, spore wall very smooth, also in SEM.

Basidiomycotina   *Cronartium asclepiadeum* (Willd.) Fries   16
   Teliomycetes
    Uredinales
     Melampsoraceae

Aeciospores orange-yellow in mass, hyaline to subhyaline by transmitted light; 22–26–(32) × 16–24 µm; globose, ellipsoid or polyhedric; a part of the surface ornamented with cylindrical warts, at a distance of 1.5 to 2 µm, a part smooth. The wall including warts 3–4 µm, without warts 2–3 µm thick. Cylindrical warts with stratified ornamentation and irregularly nodulose top.

Pucciniaceae   *Phragmidium tuberculatum* J. Mueller   17

Uredospores orange in mass, yellowish by transmitted light; 18–23 × 16–18 µm; globose to ellipsoid; wall 1 µm thick, ornamented with distinct conical spines.

Pucciniaceae Teliospores black in mass, dark brown by transmitted light; (24)–70–90 × (22)–32–36–(42) µm; cylindrical to elongate-ellipsoid; with 1 to 9 cells, mostly 5 to 7; each cell with 2 or 3 germ pores. Distal top roundish with a hyaline appendix; pedicel hyaline, as long as the teliospore or a little longer; verrucose ornamentation.

Puccinia graminis Persoon   18

Aeciospores orange in mass; yellowish by transmitted light; 14–16 µm in diameter; globose or slightly polyhedric; wall minutely warted with numerous pustulate processes.
Uredospores yellow-brown to coffee brown in mass, brownish by transmitted light; 17–40 × 13–23 µm; ovoid; ornamentation spinulose to verrucose and ocellate, the spines raising in the center of an areola, at a distance of 1.5 to 2 µm. Wall 2 µm thick. Generally 4, more rarely 3 or 5, equatorial germ pores.

Puccinia triticina Eriksson   19

Aeciospores orange in mass, subhyaline by transmitted light; 16–26 × 16–20 µm, globose or ellipsoid, wall minutely warted with numerous, small and tight processes. Uredospores rusty brown in mass, pale brown by transmitted light; 18–29 × 17–22 µm; globose or ellipsoid; ornamentation echinulate and ocellate, the spines surrounded by a torus, at a distance of 2 µm; 8 to 10 regularly distributed germ pores.

Ustilaginales   *Tilletia controversa* Kühn   20
Tilletiaceae
Chlamydospores dark greyish brown to almost black in mass, pale yellow brown by transmitted light; 18–24 µm in diameter; subglobose; reticulate, with a network 1.5–2.5 µm high and meshes of 3–6 µm across.

Ustilaginaceae   *Ustilago avenae* (Persoon) Rostrup   21

Chlamydospores dark greenish brown in mass, pale greenish brown, lighter in colour on one side by transmitted light; 4–10 µm in diameter; spherical to subspherical or ellipsoid; distinctly warty, especially on the paler side, on a finely tuberculose surface.

Ustilaginaceae	*Ustilago maydis* (DC) Corda

Chlamydospores blackish brown in mass, yellow brown by transmitted light; 7–15 µm in diameter; subglobose to ellipsoidal; wall bluntly echinulate.

Ustilago tragopogonis-pratensis (Persoon) Roussel	23

Chlamydospores purplish black in mass, pale violet by transmitted light; 12–16 µm in diameter; subglobose or ellipsoidal; delicately reticulated ornamentation with meshes of 1–2 µm across; each alveole minutely warted by some irregular tubercles.

Hymenomycetes
Agaricales
Agaricaceae	*Agaricus bitorquis* (Quélet) Saccardo	24

Basidiospores purplish brown in mass, pale reddish brown by transmitted light; 4–6–(7) × 3.5–5–(6) µm; ellipsoid to ovoid-ellipsoid; wall smooth, thick with a brown exosporium.

Amanitaceae	*Amanita fulva* Schaeffer ex Persoon	25

Basidiospores white in mass, hyaline by transmitted light; 10–13 µm in diameter; spherical with rounded hilar appendix; wall thin, non-amyloid and smooth or very minutely puncticulate.

Boletaceae	*Boletus edulis* Bulliard ex Fries	26

Basidiospores olive brown in mass, pale yellow by transmitted light; 15–19 × 5–6 µm; fusiform with supra-appendicular depression and a small subconical rounded hilar appendix; wall moderately thick, smooth or puncticulate; some guttulae in the cytoplasm.

Suillus grevillei (Klotzsch) Singer	27

Basidiospores ochre-olive to yellow-brown in mass, pale brownish yellow by transmitted light; 7–10 × 2.5–4 µm; elongate-ellipsoid to fusiform; wall smooth, composed of an endosporium, electron-dense episporium and exosporium, and an electron-translucent ectosporium.

Cortinariaceae	*Cortinarius praestans* Cordier	28

Basidiospores ferrugineous brown in mass, yellowish brown by transmitted light; 15–18 × 8–10 µm; amygdaliform or fusoid, some-

Cortinariaceae

what tapering in the distal part, with faintly delimited, proximal hilar appendix; wall, around the guttulate cytoplasm, composed of an endosporium, a smooth episporium, an irregularly thickened brownish exosporium enveloped by perisporial remnants and a thin ectosporium; ornamentation of more or less coalescent verrucae.

Inocybe mixtilis (Britzelmayr) Saccardo

29

Basidiospores sordid brown to bister in mass, pale yellowish by transmitted light; 6.7–7.2–10 × (5)–5.5–7.2 µm; ellipsoid-nodulose with distinct large rounded-conical bosses; with trapezoid proximal hilar appendix; wall moderately thickened and smooth.

Inocybe queletii Maire & Konrad

30

Basidiospores brownish ochraceous in mass, pale yellowish brown by transmitted light; 9–13 × 5–7 µm; ovoid, pruniform to somewhat reniform; with distinct hilar appendix; wall moderately thickened and smooth or almost smooth.

Inocybe subcarpta Boursier & Kühner

31

Basidiospores greyish ochraceous to bister in mass, yellowish by transmitted light; (8)–10–12 × (5.5)–6–7 µm; tuberculose-angular to subtrapezoidal with large, subconical, rounded protuberances and short hilar appendix; wall moderately thick, almost smooth.

Rozites caperata (Persoon ex Fries) Karsten

32

Basidiospores rusty brown in mass, yellowish brown by transmitted light; 13–15 × 8–10 µm; broadly ovoid to amygdaliform or limoniform, attenuated apically and with a rounded proximal hilar appendix; wall composed, around the cytoplasm, of an endosporium, a smooth episporium, a verrucose exosporium covered by perisporial remnants and a thin ectosporial pellicle.

Gomphidiaceae

Chroogomphus rutilus (Schaeffer ex Fries) O. K. Miller

33

Basidiospores olive brown to black in mass, yellow-brown by transmitted light; 17–24 × 6–8 µm; fusiform with a slight depression at the proximal part of the adaxial face; short cylindrical hilar appendix; wall thick and smooth.

Rhodophyllaceae   *Clitopilus prunulus* (Scopoli ex Fries) Kummer   34

Basidiospores sordid pink in mass, pale yellowish pink by transmitted light; $10-14 \times 5-6$ µm; ellipsoid-fusiform with mostly six longitudinal ridges which make the spores appear stellate in equatorial view; proximal hilar appendix not distinctly delimitate and spore apex smooth and rounded; wall thick and smooth, but inner layer externally longitudinally costulate.

Rhodophyllus hirtipes (Schum. ex Fries) Lange   35

Basidiospores pink in mass, yellowish incarnate by transmitted light; $10-14-(16) \times 7-9$ µm; polyhedral with slightly depressed facets; proximal hilar appendix short and truncate; wall smooth, with broadly reticulate inner layer marking the ridges of the polyhedral volume.

Pluteus atricapillus (Secr.) Sing.   36

Basidiospores pink-brown to brick red in mass, pale cinnamomeus pink by transmitted light; $7-9 \times 4-6$ µm; subglobose, obovoid to short ellipsoid; little proximal hilar appendix; wall smooth.

Russulaceae   *Lactarius lignyotus* Fries   37

Basidospores ochraceous to yellowish in mass, hyaline by transmitted light; $9-10$ µm in diameter; globose or subglobose to very shortly ellipsoid; wall amyloid in its upper zone and locally, ornamented with subhemispherical warts, pillars and more or less high crests which form a fragmentary reticulum; cytoplasm guttulate.

Lactarius pterosporus Romagnesi   38

Basidiospores deep ochre in mass, hyaline by transmitted light; $6.5-8.5 \times 6-8$ µm; globose or subglobose, with rounded-subconical proximal hilar appendix; wall amyloid in its upper zone and locally, ornamented with wing-like ridges sometimes anastomosed and surrounding numerous rounded warts and columns.

| Russulaceae | *Lactarius trivialis* Fries | 39 |

Basidiospores cream-coloured in mass, hyaline by transmitted light; $6-7.5-8.5 \times (5.5)-6-6.5$ μm; broadly ellipsoid, with rounded subconical hilar appendix; wall amyloid in its upper zone and locally, ornamented with subconical and sub-cylindrical warts forming an incomplete network of short crests.

Russula emetica (Schaeffer ex Fries) Persoon ex Fries   40

Basidiospores white in mass, hyaline by transmitted light; $(8)-9-10.5 \times 7-7.5-8.5$ μm; obovoid, with rounded subconical proximal hilar appendix; wall amyloid in its upper zone, forming an ornamentation of rounded subconical warts which are individualised or connected by lower crests in a more or less complete reticulum; supra-appendicular area irregular and warty.

Russula laurocerasi Melzer var. *fragrans* Romagnesi   41

Basidiospores pale cream-coloured in mass, hyaline by transmitted light; $8.5-9.5-(10) \times 8-8.5$ μm, with $1.5-2$ μm high ornaments; nearly globose, with a subcylindrical proximal hilar appendix; wall amyloid in its upper zone, highly ornamented with numerous rounded-cylindrical warts and simple or branched, straight or curved, festooned wing-like ridges.

Russula puellaris Fries   42

Basidospores cream-coloured to deep yellow in mass; hyaline to subhyaline by transmitted light; $6.5-9.5 \times 5.5-7$ μm; broadly ellipsoid to subglobose or obovoid, with rounded subconical proximal hilar appendix; wall amyloid in its upper zone, irregularly ornamented with warts of different height and of short ridges; almost circular supra-appendicular area.

| Strobilomycetaceae | *Strobilomyces floccopus* (Vahl in Fl. Dan. ex Fries) Karsten | 43 |

Basidiospores dark brown to brown-black in mass, reddish brown by transmitted light; $10-13-(15) \times (8.5)-9-10-(12)$ μm; subglobose to ovoid or broadly ellipsoid, with cylindrical proximal hilar appendix; reticulate ornamentation; wall thick, composed of an endosporium, smooth episporium, irregularly but completely reticu-

Strobilomycetaceae	late brown exosporium embedded in hyaline perisporium and a thin ectosporium; meshes circular to polygonal.	

Strophariaceae | *Naematoloma sublateritium* (Fries) Karsten | 44

Basidiospores fuscous to purple-brown in mass, pale fuscous by transmitted light; 6–8 × 3–5 µm; ovoid-ellipsoid with straight adaxial surface, rounded-truncated by a depressed germ-pore; distinct, small hilar appendix at the proximal part; wall thick and smooth.

Tricholomataceae | *Lepista sordida* (Fries) Singer | 45

Basidiospores pinkish in mass, sub-hyaline by transmitted light; 6–7 × 3–4 µm; ellipsoid with a little proximal hilar appendix; wall thin, verrucose by numerous pustules of various sizes.

Melanoleuca evenosa (Saccardo) Konrad | 46

Basidiospores white in mass, hyaline by transmitted light; 8.5–11 × 4–4.7 µm; oblong-ellipsoid with distinct, rounded proximal hilar appendix; wall amyloid, distinctly warty with irregular sub-hemispherical tubercles.

Oudemansiella radicata (Rehlan ex Fries) Singer | 47

Basidiospores white to cream coloured in mass, hyaline by transmitted light; 12–16 × 10–12 µm; short ellipsoid, without suprahilar depression, with distinct proximal hilar appendix; basidiospore surface minutely warted by numerous irregular tubercles.

Tricholomopsis platyphylla (Persoon ex Fries) Singer | 48

Basidiospores white in mass, hyaline by transmitted light; 7–10 × 5.5–7.7 µm; broadly ellipsoid to subglobose-ovoid with delimitate proximal hilar appendix; wall non-amyloid, thin and smooth (foveate aspect by artefact); cytoplasm with granular content.

Xeromphalina campanella (Batsch ex Fries) Kühner & Maire | 49

Basidiospores pure white in mass, hyaline by transmitted light; 6–7 × 3–4 µm; oblong-ellipsoid to subcylindrical, with distinct hilar appendix; wall amyloid, thin, appearing rugulose-plicate in SEM.

Aphyllophorales *Cantharellus cibarius* Fries
  Cantharellaceae

Basidiospores yellowish pink to pale orange or pinkish-cinnamon in mass, hyaline by transmitted light; 5–12 × 5–7 µm; obtusely ellipsoid and sometimes reniform; wall non-amyloid, thin, smooth, with an episporium, a perisporium and an outer ectosporium around the multiguttulate cytoplasm.

  Coniophoraceae *Serpula lacrymans* (Wulfen ex Fries) Karsten 51

Basidiospores deep-ochre to ochre-ustal in mass, yellow to brownish yellow by transmitted light; 9–12 × 4.5–6 µm; sub-ellipsoid, sometimes slightly depressed on the adaxial surface; wall non-amyloid, brownish yellow, 1.5 µm thick, smooth, with two layers – episporium and exosporium – beneath an ectosporium. Cytoplasm often guttulate.

  Corticiaceae *Peniophora quercina* Fries 52

Basidiospores light red, pink to carneous in mass, hyaline by transmitted light; 10–13 × 3–5 µm; allantoid with a proximal lateral hilar appendix; wall non-amyloid, thin and minutely warted by numerous tubercles, wall formed by an episporium, a perisporium and an ectosporium.

  Ganodermataceae *Ganoderma applanatum* (Persoon ex S. F. Gray) Patouillard 53

Basidiospores chocolate brown in mass, brown by transmitted light; 6.5–8.5–(9.5) × 4.5–6 µm; ellipsoid-ovate, somewhat elongate and distinctly truncated at the distal part at maturity; surface foveate and wall composed of an episporium around the cytoplasm, a thick brown exosporium with hyaline irregular cavities of a perisporium and an ectosporium.

  Gomphaceae *Gomphus clavatus* (Persoon ex Fries) S. F. Gray 54

Basidiospores ochraceous to cinnamon-buff in mass, yellowish to pale alutaceous by transmitted light; 10–14–(18) × (3.7)–4–6 –(7) µm; elongate-ellipsoid, from sub-cylindric, sub-clavate to sub-amygdaloid, with a proximal part sometimes slightly curved and a large trapezoidal hilar appendix; pustulate ornamentation; basidiospore wall composed, above the multiguttulate cytoplasm, of an endosporium, an episporium, lenticular remnants of a perisporium under the pustules of an exosporium overlined by an ectosporium.

Hydnaceae	*Hydnum repandum* L. ex Fries	55

Basidiospores cream coloured in mass, hyaline by transmitted light; 7–9 × (6)–7–7.5 µm; subglobose to broadly ellipsoid; rounded hilar appendix; ornamentation of numerous minute warts; wall thin and cytoplasm with granular contents.

Polyporaceae	*Heterobasidion annosum* (Fries) Brefeld	56

Basidiospores white in mass, hyaline by transmitted light; 4.5–6 × 3.5–4.5 µm; subglobose to broadly ellipsoid; ornamentation of irregular tuberculate warts; wall non-amyloid, thin, composed of an ornamented electron-dark exosporium, a thin perisporial layer, a fibrillous episporium and an endosporium.

	Polyporus squamosus (Hudson ex Fries) Fries	57

Basidiospores white in mass, hyaline by transmitted light; 10–14 –(16) × 4–5–(6) µm; oblong-ellipsoid to sub-cylindrical; wall non-amyloid, thin and smooth; cytoplasm containing one or several guttulae.

Thelephoraceae	*Phylacteria terrestris* (Ehrhart ex Fries) Patouillard	58

Basidiospores rusty-brown, fuscous-purple to violaceous-ferrugineous in mass, amber to pale bister-brown by transmitted light; 7–12 × 5–9 µm; ovoid to subglobose, irregularly gibbous and ornamented with subconical spines 0.5 µm high; wall composed of an exosporium and an episporium.

Gasteromycetes Lycoperdales Lycoperdaceae	*Bovista plumbea* Persoon per Persoon	59

Basidiospores reddish brown to chocolate brown in mass, brownish by transmitted light; 5–5.5 × 4.5–5 µm; globose or broadly ovoid, with a (5)–8–11 µm long pedicel which is tapering from the spore to the extremity; wall with irregularly verrucose ornamentation.

	Calvatia excipuliformis (Scopoli trans Persoon cum emend.) Perdeck	60

Basidiospores olive brown in mass, brownish by transmitted light; 4–6 µm in diameter; globose, with a 1–2 µm long cylindrical pedicel; wall with ornamentation of spool-like columns appearing locally connected.

| Lycoperdaceae | *Lycoperdon perlatum* Persoon per Persoon | 61 |

Basidiospores olive brown in mass, pale brownish by transmitted light; 3.5–4.5 µm in diameter; globose, with a 1–1.5 µm long cylindrical pedicel; wall ornamented with exosporial warts and subconical spines, covered by a perisporial-ectosporial veil.

| Deuteromycotina | *Acremonium butyri* (van Beyma) Gams | 62 |

Conidia non-septate; hyaline; 3.5–5.8 × 1.5–2.5 µm; ovoid, smooth, with slightly prominent hilum. Phialospores.

Alternaria sp.   63

Conidia solitary or in chains; mid to dark brown; obovoid, clavate or more typically rostrate, generally muriform; smooth or verruculose. Porospores. The genus differs from, e.g., *Ulocladium* and *Stemphylium* (muriform porospores) by having a geniculate, fertile (scar-bearing) beak.

Arthrinium sp.   64

Conidia solitary (hence with a unique basal hilum), unicellular; brown; 4–10 × 3–7 µm; bivalved because of a ring-like hyaline germ slit, germ slit longitudinal, hilum on or close to the slit plane; smoothwalled. Radulaspores. An additional character of the genus is the existence of highly refractive and thick septa on hyphae.

Aspergillus sp.   65

Conidia in chains, unicellular; 2–3 µm; smooth or with spines. Phialospores. Conidia very abundant in the air.

Botrytis cinerea Pers. ex. Nocca & Balb.   66

Conidia 8–14 × 6–9 µm; smooth, with basal scar. Botryoblastospores.

Briosia cubispora (Berk. & Curt.) v. Arx   67

Conidia in chains; unicellular; mid brown; 5–9 × 4–9 µm; cylindrical; smooth. Arthrospores. Conidia seceding schizogenously, sometimes possessing a circular fringe and a central plug at the level of each septum.

130

Deuteromycotina
Chalara elegans Nag Raj & Kendrick

Endoconidia (phialospores) unicellular; hyaline to light brown; 9–16 × 3–4 µm; regularly cylindrical with truncate to later rounded ends. Phialospores. Macroconidia (aleuriospores) solitary, septate; brown; 21–57 × 10–14 µm; cylindrical to claviform, more or less restricted at cross-wall level, 3–6 apical melanized cells and 1–3 basal hyaline cells. Aleuriospores. This fungus, as other species of the genus, probably exists in the air flora as phialospores.

Cladosporium cladosporioides (Fres.) De Vries 69

Conidia in chains, generally non septate; olivaceous brown; 3–7 × 2–4 µm; smooth to minutely verrucose, with truncate scar. Blastospores. Basal conidium of a chain, inserted on the top of the conidiophore, is larger (up to 30 µm), sometimes 1-septate and named ramoconidium.

Conoplea mangenotii Reisinger 70

Conidia solitary, non septate; dark brown; 7.5–11 × 6–10 µm; globose to obovoid, sometimes asymmetric; smooth with basal debris at the scar; wall structure diplothecate; longitudinal germ slit which corresponds to a less resistant zone in the wall. Radulaspores. Mature conidia at the less resistant zone appear cup-shaped with a large invagination.

Deightoniella torulosa (Syd.) M. B. Ellis 71

Conidia solitary, with 3–8 (generally 4–6) pseudosepta; brown; 48–120 × 14–28 µm; conical with rounded base; smooth, with dark prominent basal scar. Porospores. Liberation depending on air humidity linked with appearance of gas vacuoles in the conidium; the conidia are violently shot to a distance over 10 mm.

Dendryphiella vinosa (Berk. & Curt.) Reisinger 72

Conidia in branched chains, generally triseptate; melanized; 14–30 × 5–7 µm; echinulate or slightly verrucose; detached conidia may possess one basal scar (terminal conidium), one basal and one apical scar (intercalary conidium) or one basal and two apical scars. Porospores. Basal conidium of a chain sometimes with strongly melanized basal scar thickened around the pore.

Deuteromycotina

Drechslera avenae (Eidam) Sharif.

Conidia with 1–9 pseudosepta; pale yellowish to olivaceous brown; 30–150 × 11–22 µm; straight and more or less cylindrical; hilum integrated, 4–6 µm wide. Porospores. Liberated by violent jerky movements of the conidiophores due to gas vacuoles.

Drechslera spicifera (Bain.) Nicot 74

Conidia regularly with three septa; mid-brown; 20–32 × 8–12 µm; straight, cylindrical or obovoid with rounded ends; wall structure diplothecate. Porospores. The porospores exhibit a thin subhyaline area at the apex reminiscent of a germ pore, non-functional.

Fusarium oxysporum Schlecht. ex Fr. f. sp. *lycopersici* 75

Microconidia unicellular; 5–12 × 2.2–3.5 µm; straight or curved. Macroconidia generally 3–5 septate with pointed ends; hyaline; 30–66 × 3–5 µm; thin walled. Phialospores.

Geotrichum candidum Link 76

Conidia unicellular; hyaline; 6–16 × 3–6 µm; cylindrical, truncate at both ends. Arthrospores.

Gliocladium catenulatum Gilm. & Abbott 77

Conidia non-septate; light green (mature conidia); 4–7 × 3–4 µm; ovoid with rounded apex and flattened base; smooth. Phialospores.

Gliomastix murorum (Corda) Hughes 78

Conidia non-septate; grey-green; 2.5–5.5 × 2.0–4.5 µm; ovoid; smooth to rough; hilum flat and prominent. Phialospores.

Metarrhizium anisopliae (Metsch.) Sorok. 79

Conidia in chains, unicellular; 5–9 × 2.5–3.5 µm (small spores), 10–14 × 3–4 µm (large spores); cylindrical; smooth or finely ornamented. Phialospores. The species is an important insect pathogen causing green muscardine.

Nigrospora oryzae (Berk. & Br.) Petch 80

Conidia solitary, non-septate; black; 12–14 µm wide; pill-shaped; smooth. Aleuriospores.

Deuteromycotina

Paecilomyces elegans (Corda) Mason & Hughes

Conidia in chains, unicellular; hyaline; $4-6 \times 1.5-2.5$ μm; ellipsoid to fusiform; smooth with pointed apex and truncate basal hilum. Phialospores.

Penicillium sp. 82

Conidia in chains, unicellular; 3×2.5 μm. Phialospores.

Scopulariopsis brevicaulis (Sacc.) Bainier 83

Conidia in chains, unicellular; light brown; $5-8 \times 5-7$ μm; globose to ovoid; spinulose with thickened rim around the truncate base and a more or less apiculate apex. Annellospores. Liberation of conidia through rupturing of the fragile chains.

Scytalidium lignicola Pesante 84

Conidia hyaline to brown; $6-10 \times 1-3$ μm; cylindric, truncate at both ends (thin-walled), or $6-15 \times 5-10$ μm; rounded (thick-walled). Arthrospores. Melanized thick-walled, rounded conidia are reminiscent of chlamydospores; all intermediates between them and *Geotrichum*-like cylindrical, hyaline arthrospores may be found.

Stachybotrys chartarum (Ehrenb. ex Link) Hughes 85

Conidia unicellular; dark at maturity; $7-12 \times 4-6$ μm; ellipsoid; smooth or coarsely rough; hilum more or less prominent, truncate. Phialospores.

Trichocladium asperum Harz 86

Conidia solitary, predominantly 1-septate; dark brown; $15-30 \times 10-15$ μm; clavate, obovoid or ellipsoid; coarsely verrucose, with truncate base. Aleuriospores.

Trichothecium roseum (Pers.) Link ex S. F. Gray 87

Conidia in chains, 1-septate; hyaline; $12-18 \times 8-10$ μm; pyriform; smooth, with two flattened lateral scars on the basal cell. Meristem-arthrospores Conidia are blown out at the apex of the conidiophore; a basipetal chain of conidia is produced, while the conidiophore gradually shortens. Conidia liberated by rupture of the fragile chains.

Fungal Airborne Diaspores in Different Environments

Table 1. Spores in agricultural land

Botrytis cinerea
Claviceps purpurea
Erysiphe necator
Fusarium sp.
Helminthosporium maydis
H. papaveri
Ophiobolus graminis
Peronospora brassicae
P. tabacina
Puccinia graminis
P. pelargonii-zonalis
P. striiformis
Tilletia caries
Ustilago nuda

Table 2. Spores in a wheat field (Corbaz 1969)

Alternaria sp.
Chaetomium sp.
Cladosporium sp. (most common)
Didymella exitialis
Erysiphe graminis
Helminthosporium sp.
Leptosphaeria sp.
Puccinia striiformis
Sporobolomyces sp.

Table 3. Spores in vineyards (Corbaz 1972)

Alternaria sp.
Botrytis cinerea
Cladosporium sp.
Coleosporium senecionis
Leptosphaeria sp.
Plasmopara viticola
Polythrincium trifolii
Uncinula necator

Table 4. Spores in spruce stands, S. Finland, October (Kallio 1971)

Acremonium spp.
Bjerkandera adusta
Chondrostereum purpureum
Coriolus zonatus
Fomitopsis pinicola
Heterobasidion annosum
Merulius tremellosus
Peniophora pithya
Phlebia gigantea
Sistotrema brinkmannii
Stereum sanguinolentum
Trichoderma lignorum

Table 5. Spores in timber yards throughout the year, Sweden (Mathiesen-Käärik 1955)

Alternaria spp.
Aspergillus spp.
Aureobasidium pullulans
Cladosporium spp.
Mucoraceae
Penicillium spp.
Stemphylium spp.
Trichocladium spp.
Trichoderma spp.
Trichothecium spp.

Table 6. Spores in drying kilns at saw mills (Henningsson 1980)

Aspergillus fumigatus
A. niger
A. ochraceus
Paecilomyces variotii
Rhizopus rhizopodiformis
Trichoderma lignorum

Table 7. Spores around stored wood chips (Thörnqvist and Lundström 1980)

Aspergillus fumigatus
A. niger
Aureobasidium pullulans
Paecilomyces variotii
Phanerochaete chrysosporium
Penicillium piceum
Trichoderma sp.

Table 8. Spores in the air of warehouses (Moreau 1959)

Alternaria sp.
Aspergillus sp.
Botrytis sp.
Cladosporium sp.
Fusarium sp.
Mucor sp.
Penicillium sp.
Rhizopus nigricans
Trichoderma viride

Table 9. Spores in the air of bakeries (Charpin et al. 1971)

Aspergillus flavus
Cladosporium sphaerospermum
Hyalodendron album
Penicillium pallidum
Torulopsis sp.

References

Ainsworth GC, Bisby GR (1971) Dictionary of the Fungi, 6th edn. Commonwealth Mycol Inst Kew

Ainsworth GC, Sparrow K, Sussman AS (1973). The Fungi, Vol. IV A, B. Academic Press, London New York

Bassett IJ, Crompton CW, Parmelee JA (1978) An atlas of airborne pollen grains and common fungus spores of Canada. Biosyst Res Inst, Ottawa

Charpin J, Lauriol-Mallea M, Renard M, Charpin H (1971) Etude de la pollution fongique dans les boulangeries. Bull Acad Natl Méd 155:52–55

Cole GT, Samson RA (1979) Patterns of development of conidial fungi. Pitman Publ. Co., London

Corbaz R (1969) Etudes des spores fongiques captées dans l'air. I. Dans un champ de blé. Phytopath Z 66:69–79

Corbaz R (1972) Etudes des spores fongiques captées dans l'air. II. Dans un vignoble. Phytopath Z 74:318–328

Demoulin V (1968) Gastéromycètes de Belgique: Sclerodermatales, Tulostomatales, Lycoperdales. Bull Jard Bot Natl Belg 38:1–101

Dennis RWG (1978) British Ascomycetes. Cramer, Vaduz

Dimmick RL, Akers AB (1969) An introduction to experimental aerobiology. Wiley-Interscience, New York

Donk MA (1956) The generic names proposed for Hymenomycetes. 5. "Hydnaceae". Taxon 5:69–80, 95–115

Donk MA (1957) The generic names proposed for Hymenomycetes. 7. "Thelephoraceae". Taxon 6:17–28, 68–85, 106–123

Donk MA (1958) The generic names proposed for Hymenomycetes. 8. "Meruliaceae" and Cantharellus s. str. Fungus 28:7–15

Donk MA (1960) 10. The generic names proposed for Polyporaceae. Persoonia 1:173–302

Gregory HP (1973) Microbiology of the atmosphere, 2nd edn. Leonard Hill, Aylesbury

Henningsson B (1980) Thermotolerant mould on timber during kiln drying. Document IRG/WP 1109:1–10

Ingold CT (1971) Fungal spores. Their liberation and dispersal. Clarendon, Oxford

Kallio T (1971) Aerial distribution of some wood-inhabiting fungi in Finland. Acta For Fenn 115:1–17

Luttrell ES (1963) Taxonomic criteria in Helminthosporium. Mycologia 55:643–674

Luttrell ES (1964) Systematics of Helminthosporium and related genera. Mycologia 56:119–132

Mangenot F, Reisinger O (1976) Form and function of conidia as related to their development. In: The fungal spore (D. J. Webster and W. M. Hess, eds.), pp. 789–847. J. Wiley and Sons, New York London

Martin GW, Alexopoulos CJ (1969) The Myxomycetes. Univ Iowa Press Iowa City

Mathiesen-Käärik A (1955) Einige Untersuchungen über den Sporengehalt der Luft in einigen Bretterhöfen und in Stockholm. Sven Bot Tidskr 49:437–459

Moreau CL, Moreau M (1959) Pollution fongique de l'atmosphère. Sa responsabilité dans les altérations de quelques denrées alimentaires. Bull Soc Myc Fr 75:72–79

Reisinger O, Kiffer E, Mangenot F, Oláh GM (1977) Ultrastructure, cytochimie et microdissection de la paroi des hyphes et des propagules exogènes des Ascomycètes et des Basidiomycètes. Rev Mycol 41:91–117

Saccardo PA (1886) Sylloge Fungorum, 4. Pavia 1–8

Singer R (1975) The Agaricales in modern taxonomy, 3rd edn. Cramer, Vaduz

Stearn WT (1978) Botanical Latin, 2nd edn. David & Charles, Newton Abbot

Swartz D (1971) Collegiate dictionary of botany. Ronald, New York

Thörnqvist T, Lundström H (1980) Factors affecting the occurrence of fungi in fuel chips for domestic consumption. The Swed Univ Agr Sci Dept For Prod Rep R 117

Index

The references to illustrations (plate numbers) in italics, followed by page numbers to descriptions

Physiology and Biochemistry of Seeds in Relation to Germination

(In 2 volumes)

Volume 1
J.D.Bewley, M.Black

Development, Germination and Growth

1978. 122 figures, 41 tables. IX, 306 pages. ISBN 3-540-08274-3

Contents: The Structure of Seeds and Their Food Reserves. – The Legacy of Seed Matura-tion. – Imbibition, Germination, and Growth. – Biochemistry of Germination and Growth. – Mobilization of Reserves. – Control Processes in the Mobilization of Stored Reserves.

Volume 2
J.D.Bewley, M.Black

Viability, Dormancy and Environmental Control

1982. 154 figures, 72 tables. Approx. 384 pages. ISBN 3-540-11656-7

Contents: Viability and Longevity. – Dormancy. – The Release from Dormancy. – The Control of Dormancy. – Perspective on Dormancy. – Environmental Control of Germina-tion. – Author Index. – Subject Index.

The book provides a comprehensive, up-to-date treatment of selected aspects of seed phy-siology and biochemistry, including viability, longevity, dormancy, and the environmental control of germination. Discussion is at a level which will be useful to advanced students as well as to research workers and lecturers. The subject is treated in a critical manner, ensuring that areas of ignorance, over-assumption or incomplete knowledge are fully dis-cussed, and modern developments, discoveries and concepts receive thorough considera-tion. Together with volume 1, this book represents the most wide-ranging advanced text available in this subject area.

From the reviews of volume 1:

"…will surely become the standard work on seed physiology for the next few years."
Times Higher Education Supplement

"…but there remains a need for a text that gives a comprehensive view of the whole.This need has now been met by Bewley and Black in a two-volume compendium."
Plant, Cell and Environment

Basidium and Basidiocarp

Evolution, Cytology, Function and Development

Editors: **K.Wells, E.K.Wells**
With contributions by numerous experts
1982. 117 figures. XI, 187 pages. (Springer Series in Microbiology). ISBN 3-540-90631-2

Contents: Introduction. – The Significance of the Morphology of the Basidium in the Phy-logeny of Basidiomycetes. – Ultrastructure and Cytochemistry of Basidial and Basidiospore Development. – Meiotic Divisions in the Basidium. – Replication of Deoxyribonucleic Acid and Crossing Over in *Coprinus*. – Biochemical and Genetic Studies on the Initial Events of Fruitbody Formation. – Control of Stipe Elongation by the Pileus and Mycelium in Fruitbodies of *Flammulina velutipes* and other Agaricales. – Metabolic Control of Fruit-body Morphogenesis in *Coprinus cinereus*. – Author Index. – Subject Index.

Springer-Verlag
Berlin
Heidelberg
New York